Science and
Writing Connections

Science and Writing Connections

Robin Lee Harris Freedman, Ph.D

Dale Seymour Publications®

Dedication

*To Mom and Dad, for the unqualified love and infinite patience
you have always given your youngest daughter.*

Managing Editor Cathy Anderson
Senior Editor Jeri Hayes
Production/Manufacturing Director Janet Yearian
Production Coordinator Joan Lee
Design Manager Jeff Kelly
Text Design Don Taka
Cover Design Don Taka
Composition Joe Conte
Cover Photos Comstock, Inc.

ISBN 1-57232-576-3
Printed in the United States of America
5 6 7 8 9 10 11   06 05 04 03 02

1-800-321-3106
www.pearsonlearning.com

Contents

To the Teacher

*T*he purpose of this book is to supply you, the classroom teacher, with activities and strategies that

- use higher-order cognitive skills in the expression of conceptual understanding
- accommodate various learning styles
- allow you to teach and assess students' abilities to explore science issues, ideas and concepts
- promote the pacing and repetition important to the learning processes advocated in this book, since brushing by many ideas in a superficial manner will not create thoughtful learners

If you teach in a discipline other than science, the science-centered examples and samples in this book can easily be adapted to your needs. The need for conceptual understanding and literacy is not exclusive to science classrooms. Another purpose of this book is to advocate student-centered classrooms where

- science is studied by asking questions, by investigating, and by making the content relevant and personal to students
- students explore in-depth issues, ideas, and concepts that are relevant to many cultures, countries, and our world

A word about the role of nurturing in a student-centered classroom. Thoughtful, curious, enthusiastic learners need nurturing—that is unless you teach in lower elementary classes where such enthusiasm and curiosity is the norm. Nurturing your students should be a part of your management plan. It includes developing an atmosphere of trust and mutual respect. Each of you does this in accordance with your personal teaching style. Thoughtful learners need time to think and question the environment and events that take place around them. The classroom can be a place for thoughtful questioning and investigation of issues, ideas, and concepts. This type of classroom excludes racing through content in an attempt to attain greater coverage. Curiosity abounds in an atmosphere where negative criticism is minimized and where all student responses are validated.

The activities in this book take time. The following suggestions may help you feel comfortable using them in your classroom:

- Familiarize yourself with the activities before trying them out with your students.
- Try different activities with different classes.
- Ask your students for feedback on how the activity worked or did not work.
- Collaborate with other teachers on obvious multidiscipline activities, such as skits, stories, and comics.

Students need practice with new techniques before you can expect them to demonstrate achievement on assessments. Processes are harder to acquire than isolated facts. If your students are engaged emotionally with content and can find patterns they understand, they learn information faster and retain it longer. Research indicates that it takes twenty-five repetitions before the brain creates new neural pathways. However, if your students are emotionally involved, it may take only once! Enthusiasm and excitement are emotions you want to foster in your classroom.

Besides repeating activities using different science content, consider collaborating on activities with your colleagues, especially those that are long-term or that take several iterations before completion. Language arts and English teachers will welcome the use of science content as a basis for their writing activities. Math teachers will gladly cooperate on research projects and graphing exercises. Social studies teachers will help you set up historical cultures that coincide with historical science figures and events. Approach your colleagues early in the school year and plan joint activities with them. The time and effort required will be very worthwhile.

Within each chapter of this book, activities are arranged from less difficult to more difficult in regard to thinking skills. As you progress through units of instruction, you might start with the easier activities and then move to the more difficult ones. The developmental abilities of your students should guide your choices. Use the indexes and the table of contents to locate activities that fit your purposes—whether they are learning or assessment activities.

To return to the matter of enthusiasm in the classroom—enthusiasm is contagious. It starts with your modeling this behavior toward the issues, ideas, and concepts of science. It continues when you make Science and Writing Connections between content and your students' lives. When students develop positive attitudes towards the processes and concepts of science, you will be able to develop thoughtful, curious, enthusiastic learners. It's not easy, but it is rewarding.

Acknowledgments

*I*n the process of writing both editions of this book, I have had the support and inspiration of many people. I am grateful to my husband and my family, who have encouraged me in all my professional endeavors. Thank you to my colleagues from near and far, who have tried out the activities in the first edition with their students and let me know how things worked—or didn't! Thank you to the students whose writing I quote here with their permission.

Without a publisher and editors this second edition of *Science and Writing Connections* would only be a dream. I thank Linda Roach for her editing talents. As a former middle-school science teacher, her questions helped focus the work. Thanks to John Lanyi for asking thought-provoking questions and encouraging me to think differently about this edition. Addison Wesley Longman once again has put their trust in my abilities; for this, I am ever grateful. Thank you to Angelica and Sasha Freedman for their help in the word processing of this manuscript.

Finally, I would like to thank you, the purchaser of this book for the willingness to change and enhance your current teaching practices. Enjoy, adapt the activities in this book to your needs, and know that you help your students on their journey to scientific literacy.

Robin Lee Harris Freedman, Ph.D
January 1, 1998

Process Writing in Science Classrooms

1

*T*his chapter introduces the art of process writing using one model. Process writing has been advocated by the National Writing Project (NWP). Graduates and others of NWP institutes and those who have been mentored by graduates have successfully taught their students process writing. Just like the many variations of the scientific method of experimentation, the processes used in writing can be taught and used in science classrooms. In fact, writing—especially creative writing—can enhance learning in science classrooms. Process writing activities in science classrooms can help students achieve conceptual understanding at higher cognitive levels. Higher-level understanding leads toward literacy, including scientific literacy, which is a major emphasis in current education reform.

This chapter discusses scientific literacy, the way the writing process is used in a classroom, how writing generated by activities in this book can be used in assessment and become part of a student portfolio, and how teachers can use writing in their own action research projects.

Concepts and strategies presented in this chapter include:

- Scientific Literacy
- Writing as a Process
    - Putting It Together
    - Response Groups
    - Teacher Response
    - Sharing/Publishing Student Writing
    - Scoring Student Writing
- Portfolios
- Action Research

When we try to pick out anything by itself, we find it hitched to everything else in the universe.

John Muir

Scientific Literacy

Recognizing the changing needs of society, the National Research Council (NRC), the National Science Foundation (NFS), and other reform projects advocate that science instruction should not be satisfied with talking about science—the transfer of a specific core of knowledge—as was traditional for the past forty years—but that students actually do science. Scientific literacy in the past meant memorizing a body of facts and concepts. But that isn't enough any more. In order to be successful today, students also must know how to actually engage in scientific inquiry. They must also know the history and nature of science, and understand how science connects both to their own lives and to the society in which they live. (National Research Council, 1996) Students can only learn these things by actually asking questions, conducting investigations, and thinking about what they have discovered.

As a result of this need for increased science understanding, the NRC, with the help of educators, scientists, and teachers, has produced a new set of science standards, the National Science Education Standards. (National Research Council 1996) These standards describe new learning outcomes for students in science classes. Teachers need to broaden the types of strategies they use in order to teach and assess student achievement in alignment with these new learning outcomes.

Recent teaching models, based on constructivist theory, have stressed the need to move away from the factory method of production as a model for school organization. This method worked fine when just knowing the facts was enough. But it does not give today's students the tools they need to work in the next century. Inquiry, not memorization of core knowledge, should be the focus of science instruction. Instruction should be student-centered and revolve around local instruction and societal issues. Direct instruction must give way to practices in which higher-order thinking skills and science processes are used. Students need a learning environment where they have opportunities to think, rethink, explore, fail, experiment, gather evidence, perform, and take responsibility for their own learning. Using constructivist learning models is a move in this direction. (Brooks and Brooks 1993)

Constructivism is a psychology of learning. It is based on the premise that the learner has to construct his/her own knowledge. It refutes the idea that learners are a "tabula rasa" and embraces the idea that knowledge is individually constructed. A second premise is that the construction of knowledge is facilitated by social interaction. (Brooks and Brooks 1998) In the application of constructivist theory, teachers need to question students for their prior knowledge regarding science concepts. Most always, students come to class with naïve conceptions about how the world works. These naïve conceptions, ones that are not based on scientific knowledge, are what students have used to explain scientific phenomenon that they have experienced. For example, students may think that the phases of the moon are caused by the earth's shadow, or that the seasons are caused by the distance we are from the sun at any point in our orbit. These are naïve but workable conceptions that may stay with students all their lives.

Brain research data indicates that the brain is constantly changing as it grows and interacts with the environment. (Caine and Caine 1991) New research dispels the recently held notion that the brain is two-sided. It is not that simple. Scientists now acknowledge that each person's brain is unique, which makes the ways individuals learn unique. In fact, Howard Gardner recognized this when he identified, in 1982, the following seven different intelligences: linguistic, spatial, musical, logical/mathematical, body/kinesthetic, and interpersonal and intrapersonal social intelligence. (Gardner 1982; Marks-Tarlow 1996) Since students learn differently and change the way they learn depending on many factors, it follows that they need different ways to express their achievement.

The creative writing activities in this book offer teachers many tools to assess these different expressions of achievement. The variety of activities allows teachers to develop and assess student understanding at higher cognitive levels. They allow students to express their understanding in many different ways. The activities are the means by which students can think about what they do and do not know and reflect on their personal growth. They are activities that can be used for learning and assessment, and for teaching students how to think and write at higher cognitive levels. Higher-level thinking takes time and practice, which may not work with the amount of content usually taught. The NRC advocates teaching less content in favor of greater comprehension. The activities in this book will help you do that.

Scantron-graded exams will have to give way to thoughtful examination of student work—by the teacher and the student! We need to teach our students how to find knowledge, how to discern between what is opinion and what is fact, and how to express themselves in a coherent manner. We no longer have to fill them up with content facts. They will seek out the answers to their own questions. They will seek to increase their own understanding of science concepts. You can create lifelong learners who will become scientifically literate in their own time. You are a teacher; you can accomplish miracles. It just takes time and hard work. The activities in this book will help you.

Putting It Together

The creative writing process used in this book is made of up four interactive, recursive stages.

- Prewriting evokes ideas and gives insight into students' prior knowledge.
- Composing is the written expression of those ideas.
- Revising allows students to focus their ideas so that the written presentation is logical and valid.
- Sharing gives students an outlet for expressing their ideas and an evaluation of their skill as a writer.

Teachers can use writing produced at any stage of the process to assess student understanding and achievement. In fact, you can follow a student's growth as a writer and a scientist by collecting several different kinds of writing into a student portfolio.

It is important to recognize that the writing process is not linear, but recursive and interactive. Recursion occurs as each stage of writing becomes an integral part of the other stages, and the whole process folds back on itself many times as a writer works through a piece. Interaction occurs when parts of the process feed on each other. The process follows the strategy of spiraling, each step overlapping and building on previous experiences—surprisingly similar to other models of learning. Figure 1-1 is a cluster of the writing process, showing how the stages are related.

Figure 1-1
The Writing
Process

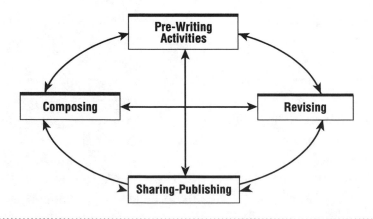

Constructivist learning models have a stage called stabilization. During this stage, students share their new ides with others, do further experimentation and research, and construct new knowledge. When you use the complete writing process from prewriting to publishing, you help students construct new knowledge. Instead of merely taking a quick look at science concepts, students spend time thinking about them. When students think about concepts and write about them, they are more likely to internalize them. Students need practice writing spontaneously. They also need to write carefully and revise many times, until their writing is a polished product—writing that reflects understanding, writing that is shared proudly.

Example A concept that has been thoroughly explored using the writing process might unfold as follows. (This example was modified from a story shared by an eighth-grade science teacher.)

Prewriting Students brainstorm about weather. They bring in an article relating to one of their top five ideas—weather prediction. They evaluate the article for naïve conceptions or myths. Students research the science concepts that are the basis for weather predictions. They design an assessment checklist that will be used during response groups and for final evaluation.

Composing Students write a rough draft of a "Dispelling Weather Myths" paper in the following form: the myth, the science behind it, experiments you can do to test the myth against the science, and what we found out.

Revising Students share their papers in response groups. Using their checklist as a guide, students give responses to each other's papers. Students revise their work based on peer responses. The teacher uses the checklist to give responses to student writing. Students self-edit and rewrite their papers.

Publishing/Sharing Students hold a weather fair. They set up their experiments, make a poster of their results and conclusions as a backdrop to the experiments, and then have a "share-a-thon." Students who researched the same weather myths then share information in smaller groups.

Response Groups

Response groups are small groups of three to five students who read their writing to each other for the purpose of obtaining constructive evaluation. Language arts teachers who use response groups report that these small, interactive groups free the teacher from being the sole authority, help students discover a better sense of audience, reduce teachers' paper load, motivate

Editing is easy. All you have to do is cross out the wrong words.

Mark Twain

students to revise papers, expose students to a variety of writing styles, and build a sense of community. Response groups also improve critical thinking processes, increase the amount of revision students undertake, and reduce apprehension about writing. (Shoemaker 1986)

Cooperative learning is not a new idea. But peer review of writing is a special form of teamwork; students need to know that these groups are different from the teams they use for problem solving or laboratory work. In response groups, students work together to build trust and respect for each other's writing and for each other. Using a set procedure, they evaluate each other's work. Group members help each other delete unnecessary information, add useful information, and write papers that stay focused. The purpose of a response group is not to write or rewrite papers for one another. Group members should motivate one another to do well and to improve their individual writing. As students work cooperatively, they give each other feedback on how their writing is progressing.

Students need supervision and direction the first few times they work in response groups. The first time you ask students to respond to the work of other students, they may not stay on task as well as they will in the future. Be patient and encourage students to help one another. You might want to designate the first response groups, perhaps with a strong writer in each group.

When working in response groups, students should be quiet listeners while a writer is reading his work. Then listeners can follow with positive suggestions. No put-downs should be allowed. It is important that students learn how to comment constructively. Before doing a group exercise, have small groups of students make lists of negative and supportive ways of saying the same thing. Share these lists with the class, and have everyone copy them into their daybooks. (See Science Daybooks, Chapter 3) At first, students probably will laugh when they recognize the polite phrase for "This is bad," but soon, making constructive comments will come naturally. Notice that the constructive examples in Figure 1-2 are all in the form of questions that ask for specific information of the writer.

Negative Criticism	Constructive Criticism
"It stinks."	"Could you be more concise in this paragraph?"
"Lousy grammar."	"Do the subject and the verb agree in each sentence?"
"You're confused."	"Could your statements be more clearly written?"
"That's a dumb question."	"What is the main question you are asking?"

Figure 1-2

Using Constructive Criticism

The writer should ask specific questions of the group. She might ask the following types of questions: What is the most interesting part of my paper? Did I keep to the topic? Do you understand what I wrote? If not, where did you get lost? Did I use science content correctly? Remind students that the comments they hear are suggestions only. Students ask for advice, but they don't have to use it.

Response groups are not for every student. You will have those few who are completely intrapersonal in their social learning style. (Marks-Tarlow 1996) Don't force those individuals to work in a group. They will do better working by themselves with your feedback. Perhaps through time and observation of others' success, these students might ask to join a response group.

Example Here is one response group format. (Adapted from Gere and Abbott 1985)

1. The group divides the amount of time allotted for the activity into the number of members of the group. Each writer is allotted an equal fraction of the time. One member of the group is designated as timekeeper.
2. The first writer reads his selection aloud twice, taking a short break between readings.
3. The writer does not comment or apologize for the selection.
4. Listeners make notes between readings and during the second reading, but not during the first.
5. Each listener offers oral and written comments on the selection according to a predetermined list of criteria. The writer notes all comments.
6. Each remaining member of the group reads his or her selection and receives comments during the time allotted.

Teacher Response

The writing process may seem to run without teacher input, but actually you are an active member. Your main purpose is to ease students through the process with your help and understanding. You should be responding or conferencing with students on an ongoing basis. Long before students' work is scored, you should have an idea of what they are writing about and guide them past naïve conceptions or pitfalls that you know await them.

Teacher responses are typically the major responses that students receive in science classes. The teacher is "the authority." When students use response groups and write for other audiences, your role shifts. You become a facilitator, a consultant, or an advisor. In this capacity, you can circulate around the room providing encouragement, positive comments, and suggestions for improvement. You can ask questions that challenge and encourage students. Oral comments now save time later. Before a student is set in a naïve conception, you can help her correct the problem. Watch out for the pitfall of becoming the style and usage editor. Allow students to correct their own errors. When you act as consultant, you transfer ownership and responsibility of student work to students themselves.

Many students want to share their work with others, although some may be shy at first. A classroom built on trust and mutual respect will bring out the shy ones. Encouragement goes a long way, too. This approach can work in a science classroom. Rather than writing to "the authority," students are writing for themselves, using their personal learning-style strengths.

When you take on the role of evaluator after assignments are handed in, remember your earlier role as facilitator. Score work according to a predetermined scoring guide, preferably one you and your students have designed together and discussed. First, score content comprehension. Naïve conceptions should be noted and discussed with the whole class. Next, look at how style and usage enhances or distracts from your ability to interpret the assignment. If you make individual corrections, give extended notes on no more than three of these corrections. Start with positive comments and then list strategies for improvement. Ask students to correct the three noted errors for their next assignment. Have them keep a page in their daybooks for style and usage notes. You might also make a master list of common style and usage errors for the classroom. Ask volunteers to make a list or chart of correct style and usage samples. Post these samples in the classroom. Better still, work with the language arts teacher to correct errors.

Sharing/Publishing Student Writing

Sharing their writing helps students realize that their work has value. The format for student sharing can range from a simple paragraph that is read aloud in class to an entry in a national writing competition. Students can share their work in a fair of some kind, a cross-grade tutoring project, an open house, or a classroom or school display. Such publishing opportunities broaden the audience of student work beyond a single teacher, "the authority." Sufficient time should be designated to the production of work destined for publishing. If desired, writing could be published voluntarily and anonymously. However, the challenge of ownership adds a new dimension to student writing and builds self-esteem. Recognition helps tie the writing to other aspects of a student's life.

Other ways in which students can share their work include having bilingual students translate their writing to another language; having students write materials for English as a Second Language (ESL) and Limited English Proficiency (LEP) students; creating a class or team magazine or newspaper; adapting work for a play, skit, or production to be shared with another grade level; and entering competitions or fairs.

Cross-Age Sharing

One teacher encouraged his students to share their science writing with younger students. His students divided into small groups and visited an elementary school, giving demonstrations and doing experiments they had previously written up. It took coordination and hard work, but the increase in student interest made it worthwhile. The teacher won a cash award from National Science Teachers Association (NSTA) for his project. With that award, he paid for a NSTA convention trip for all the teachers in his department.

Setting up a similar program in your district is worth the effort and has several benefits. For middle- or high-school students to share their knowledge with younger students, they must thoroughly understand the concepts and processes. Sharing information helps reinforce learning and is part of the constructivist model of teaching. Self-esteem is enhanced when students take on the role of teacher and find they can handle it. There is nothing like the curiosity and enthusiasm of a seven-year-old to cure even the hard-core uninvolved older student. Other benefits include providing an opportunity of success for participating students, increasing interest in science courses, and opening communication between different grade levels.

A cross-age sharing program works best when many people are involved to share responsibilities. Besides including your colleagues, enlist the aid of parents and other volunteers. Many of the writing activities in this book produce possible sharing items—for example, songs, skits, projects, demonstrations, stories, myths, naïve conceptions, comics, and investigations.

Using Bulletin Boards

Bulletin boards are a quick way to share and publish student work and to inform others about your curriculum. Any work your students produce can be shared on a bulletin board. You set the standard and the way students will react to posted work. The who-or-what-am-I? activities in Chapter 6 make interesting bulletin boards. Bulletin boards where students can interact are also desirable. Science Scope, a NSTA publication, has a bulletin board column and annual competitions for exciting and sharable boards. Figure 1-3 gives some tips for maintaining bulletin boards in the classroom.

Interactive bulletin boards make the sharing/publishing of student work part of classroom learning. For example, students in one class wrote one-page summaries of an imaginary project proposal asking National Science Foundation (NSF) for funds to do research on a local issue. In the summaries, students listed potential questions that they wanted answered. Students from other classes read the proposals, responded to the questions, and chose the proposal they would fund. Students put these responses into a box by the bulletin board. At the end of two weeks, the responses were tallied and the winning proposal was honored.

**Figure 1-3
Tips for
Maintaining
Classroom
Bulletin
Boards**

- Change boards often.
- Put students in charge of creating titles and boards that complement current work.
- Use a sign-up sheet for bulletin board maintenance.
- Let the boards evolve along with your science units.
- Set up parts of a board for current topics that will encourage classroom discussion and provide ideas for writing assignments.
- Include photos of your students.

....

Sharing on the Internet

Many places on the Internet provide opportunities for students to share work in progress, publish articles, or contribute data to other student-centered research projects. Students can design and maintain their own Web page. For starter ideas, if you are connected to the Internet, try out the National Geographic KidNet, or NASA's educational section. Web sites change faster than adolescents, but if you use one of the current search engines included with any Internet interface program, you will find many places where your students will be able to publish their work and share ideas.

Scoring Student Writing

Using different writing strategies enhances students' ability to communicate their scientific literacy and increases their written communication skills. Using different writing activities for assessment allows students with strengths in different learning styles to shine in their areas. With these two goals in mind, you should assess both the content of your students' writing and their ability to express their knowledge in written form. However, when you assess student writing, be sure you are evaluating those skills that are appropriate to the task. Research indicates that red ink evokes a turnoff in student response. A returned paper, heavy with corrections, will have little influence on improving student writing. (Emig 1971) Besides, you are using these activities to assess your students' knowledge in the fields of science, not to see how well they can write. Leave the red ink to your colleagues in the language arts department—better yet, collaborate with them.

Evaluation will be with us always.

Grace M. Burton

It is not necessary to grade everything. You may believe that every student paper needs grading. It doesn't. The important task is to check for science concept understanding. You can use several alternatives in place of copious amounts of red ink to evaluate science writing in the science classroom. You might try giving comments via cassette tapes or on postcards, having students pre-evaluate their own work, having students develop checklists for self-evaluation, using rubrics, and having students evaluate each other's work. Each of these strategies will be discussed in more detail later in this chapter.

When you respond to student work, start your comments with a positive statement. Then give your comments for improvement. Postcards, because of their size, help keep your comments concise and to the point. Create a classroom drop-box or post office where you can "mail" your comments to students. Or use cassette tapes to minimize writing time.

Self-Evaluations

You can shift some of the responsibility for evaluating student work to students themselves. Students can self-evaluate their work by filling out an evaluation form and attaching it to their work. Self-evaluation forms should be used when students are submitting final work that has gone through several drafts.

Criticism can merely reflect a difference in perceptions.

Anon.

For example, a self-evaluation form for a song written by students might look like the example in Figure 1-4.

Figure 1-4
Sample Self-Evaluation Form
(based on Weather brainstorming in Chapter 2)

Name _____ Date _____

Topic _____

1. How did this song show what I know about how weather instruments work?
2. Which weather instrument did I describe most scientifically? Explain.
3. For which weather instrument could I write a better description? How could I improve it?
4. Did all of my stanzas rhyme? (student's choice)

Directions for developing self-evaluation forms:

1. Have students brainstorm what they think are the most important items they should learn/know/be able to do for the current topic/issue/idea. Make sure that your ideas are part of the brainstorming.
2. List these ideas on the board.
3. Prioritize the items from the most important to the least important.
4. Cut the list to the top three items. Save the others for another time.
5. Restate the top three items in question form.
6. Allow each student to choose one other item from the list that he or she feels is personally important.
7. Duplicate the top three chosen items for the whole class. Allow room for the fourth student-chosen item on the page.
8. Distribute the form to the class.

Checklists

A checklist is a list of essential components for a writing assignment. Checklists provide a quick and simple means of scoring student work. They can also be used to document student participation. One sample checklist is shown in Figure 1-5.

Using student-developed checklists teaches students to take responsibility for their work and provides a focus for student self-evaluation. You may wish to use a sample checklist the first time students use a checklist to score their own work. Tell students that the next time they write, they will be given the responsibility of producing the checklist.

Be flexible in using checklists. One teacher found that after students worked on a project, they wanted to re-evaluate the point distribution. The classroom debate was lively, the arguments were well presented, and the class voted with the teacher's approval to alter the point distribution. Checklists can be used as a form of instant feedback on work in progress. You can do a quick check of a project or piece of writing, return it to the student or group, and they can use the checklist as a guide for improving their work and their score.

Name _____ Date _____

Item	Completed (check when completed)	Points (100 total)
1. Completed prewriting. (State type.)		10
2. Wrote opinion sentence.		5
3. Made list of researched facts.		15
4. Supported opinion with examples.		10
5. Wrote rough draft.		20
6. Had peers review work.		15
7. Edited for grammar.		10
8. Rewrote draft.		15

Comments on this paragraph:

**Figure 1-5
Sample Student Checklist**
(for a revised paragraph on Fact and Opinion, Chapter 4)

Directions for creating student checklists:

1. Have students brainstorm what they think are the most important components of the assignment. Write the list on the chalkboard.
2. In small groups, have students prioritize the ideas. Have students distribute points for each item on the list. It works easiest if you start with 100 points.
3. Discuss each group's work. From this discussion, create a class checklist with point distribution.
4. Duplicate checklists for students to use as a cover sheet for their assignments.

Rubrics

When you use different forms of creative writing, you expect a variety of responses from your students. (Freedman 1994a; Freedman 1994b) As a result, a broader style of scoring is needed than that which traditionally has been used to score papers. A rubric works well for this purpose. A rubric is a set of predetermined standards that is divided into different levels of achievement. Letter grades, percentages, or points may be awarded for these levels. Rubrics may be developed for specific assignments. As with checklists, students should be allowed to participate in the creation of rubrics. You can see an example of a rubric with a five-level scoring guide in Figure 1-6.

Figure 1-6
Sample Five-Level Rubric (for a written assignment)

	Achievement Level	Paper has all or most of these characteristics:
4	Exemplary	• concise, well organized, consistent, complete • reader does not have to guess at writer's intent • precise, varied vocabulary
3	Standard	• well organized • generally clear intent • sufficient detail
2	Apprentice	• inconsistent, acceptable organization, lacks details, basic idea present • does not have clear intent • vocabulary imprecise
1	Novice	• not adequately developed • generic, hasty, indirect reference to assignment • little or no detail
0	No achievement	• paper is blank, unreadable, or does not address topic

Reprinted with permission from *Open-ended Questioning: A Handbook for Educators*, Addison-Wesley Publishing Company, 1994.

This style of scoring works to help you sort out your students' responses to different creative writing activities. Once a piece of writing is evaluated using a rubric, give students the option of rewriting their papers for a higher grade. If students are satisfied with their initial score, the work is finished. Allowing revisions removes the anxiety of writing perfectly the first time.

Scoring guides may be used analytically or holistically. An analytical approach divides the assessment and awarding of scores into components. Each response to a section of the rubric receives a score. The completed paper receives a total score composed of individual scores. A holistic approach evaluates the total work as a blend of its individual parts. Students receive one grade for their response. In the scoring guide in Figure 1-6, the characteristics might be graded analytically or holistically.

Designing a rubric is as simple as organizing a topic or concept objectives into a logical manner on a grid. Decide which components of student knowledge you plan to assess. The type of components will depend on the kind of writing you are assessing. Focus questions can help you determine the various components of student knowledge. Figure 1-7 provides one example of the types of questions you might ask for four different components of student learning.

**Figure 1-7
Guiding Questions for Setting Up a Rubric**

Components	Questions
Conceptual Understanding	What are the big ideas?
Content Knowledge	What are the facts, illustrations, descriptions, and examples the student should have internalized while studying the big idea?
Critical-Thinking Processes	What kinds of thinking processes is the student engaged in while responding to this activity?
Communication Skills	How well can the student communicate? Has she paid attention to grammar, style, and usage in written communication methods? Has the student presented knowledge with competence if he is using a visual or oral method of presentation?

Reprinted with permission from *Open-ended Questioning: A Handbook for Educators*, Addison-Wesley Publishing Company, 1994.

Once you have organized the objectives, decide on the number of score levels you will acknowledge. Keep in mind that rubrics can raise or lower self-esteem. The headings you use to separate score levels can make or break a student's enthusiasm. Using Exemplary, Standard, Apprentice, and Novice level headings is one effective approach. Next set the criteria for score levels. After using several scoring guides you designed, your students will be willing and able to help you "fill in the grid" for a particular assignment.

The scoring guide shown in Figure 1-8 was used to score a research project. Students defined the content section of the scoring guide and the qualities of a standard presentation before the work was due. Students received scores for each area of the scoring guide, making this an analytical rubric. Scores were weighted. The content section was worth three-fifths of the total grade.

Figure 1-8 Sample Rubric (for a science research project)

Achievement Level	Content	Process	Presentation
Exemplary	• shows expertise for subject area according to classroom notes • written in own writing, not copied • interesting, engaging writing • shows reliability	• report is more than ten pages • ink or typed • title page present • bibliography present and in proper format • author page present • pictures present • no grammar, style, or spelling errors	• presentation is exceptional • visual aids excellent • engages audience in subject with some type of activity
Standard	• shows expertise for subject area according to classroom notes • written in own writing, not copied • interesting, engaging writing	• report is ten or more pages long • ink or typed • title page present • bibliography present and in proper format • author page present • pictures present • few grammar, style, and spelling errors	• student shows competence with subject • answers questions • uses visual aids–laser disk, pictures or other
Apprentice	• written in own writing, not copied • shows competence in subject area according to classroom notes • interesting, engaging writing	• report is ten pages long • in ink • title page present • bibliography present • author page present • pictures present • few grammar, style, and spelling errors	• student shows competence with subject • answers questions • doesn't use laser disk
Novice	• written in own writing, not copied • shows some competence in subject area according to classroom notes	• report is ten pages long • in ink • title page present • bibliography present • author page present • pictures present • few grammar, style, and spelling errors	• no questions answered • no visual aids
No achievement	• does not turn in report	• no report	• doesn't give an oral presentation

Portfolios

Portfolios can play a significant assessment role in a science class. Many of the writing activities in this book can be used as exemplars of students' abilities in science inquiry and conceptual understanding. This section explores several aspects of portfolios—their role and characteristics, what research says about them, and how to develop them in the science classroom.

Portfolios have many roles within a school community. In many middle schools across the nation, education goals have changed. Schools are experimenting with student-centered learning. Teachers are forming interdisciplinary teams and writing an integrated curriculum that is developmentally appropriate for their school population. This style of teaching needs new tools for student assessment. A portfolio is one tool that can be used efficiently to record student learning in this new learning environment.

Benefits of Portfolios

Portfolios play an important role in many classrooms, but teachers who use them disagree about what a portfolio is and what its purpose is in the science classroom. However, most will agree that a portfolio:

- is organized collection of student work
- is not tests
- focuses on personal, meaningful goals
- allows for reflection
- acknowledges that a single assessment item does not adequately measure ability
- can be an excellent reference or resource
- documents efforts
- can be an adequate picture of the abilities of a student
- can be an accurate, graphic, simple accountability system
- offers a place to reflect on learning
- provides the opportunity to revisit earlier work
- is a tool to show students how to organize work

Research on the use of portfolios supports the idea that portfolios appear to

- encourage sustained effort and improvement
- support higher-order thinking skills
- encourage student ownership
- invite diversity

- document efforts
- preserves multiple perspectives of student learning
- aid students in becoming better learners
- serve as a tool for promoting students' understanding of themselves as learners

Developing portfolios provides students with numerous learning opportunities.

- Students are encouraged to revise their work.
- Students are responsible for choosing items that will go into the portfolio.
- Students can review many aspects of learning.
- Students can see how much work they have completed.
- A portfolio can function as an organizing tool.
- Students see what they are able to do.
- Students become evaluators and critics of their own work.
- Students collaborate, rather than compete, with others to produce the best evidence of personal learning.
- Portfolios promote self-confidence.
- Students can develop individual talents.
- Students use personal judgment.

Storing Portfolios

The amount of space you need to store materials while the portfolio is in progress depends on the age level of your students and the type of portfolio you have students construct. It also depends on what you have the students save. The minimum amount of space you will need is a hanging file folder box with a folder for each student. Some teachers have wooden shelves made; others use boxes or other space. One teacher used technology to help him store his students' work. Students used one class a week in the computer lab word-processing information that was stored on individual disks.

Managing Portfolios

This brings us to time management. How often do you want students to update, or work with, their portfolios? That depends. Once you've determined the purpose of using portfolios, you can decide how much time you want students to spend on them. Be flexible—you are teaching a process to your students. They will need time and plenty of help organizing their work. They also will need time to reflect on their work; decide if this will be done during classroom time or as homework. The age of your students and the amount of experience they have had with portfolios will determine in part how much class time they will need. If this is your first time using portfolios, you may want to keep one too! If you think portfolios are important, give students plenty of class time to allow success with the development of a personal portfolio.

What Goes into a Portfolio?

At first you may have students save all their work. When students do their first gleaning, they will see just how much work they have done. Often, students forget their work as soon as it is completed. When they sit down with their filled folder, it is a wonderful reality moment. You probably will hear comments such as these: "I did all this?" "Did you put someone else's work with mine?" "There sure is a lot here!"

Scoring Portfolios

Using portfolios in your classroom requires a change in attitude toward traditional assessment practices. Teachers using portfolios give up the control that grading every paper provides. In fact, a completed portfolio does not have to be scored at all. You can grade the process it takes to put a portfolio together, but chances are, most of the individual samples that go into a portfolio have been scored already. Students become partners with you in the grading process.

How different is a portfolio from a teacher's grade book scores? A portfolio consists of multiple samples of a student's work over time. Some samples carry more weight than others, but all are a reflection of the student's learning. The difference is in the reflection—the student makes judgments on his own work and decides what to include as evidence of his progress. Instead of scores in a grade book, examples of work come alive in a portfolio. Each example is chosen by the student to represent a competency in a particular science process or in conceptual understanding. The shift is from teacher-directed to student-choice. You don't lose anything in this shift. Most work that will be chosen for the final portfolio has already been through the scoring mill. And because the work has already been scored, you can focus on the process of building the reference—the tool—not on scoring papers again.

Other Considerations

In the process of developing a portfolio, students need to know that they have ownership of their work. Initially, you may prescribe that all their work goes into the folder. But when it comes time to showcase or build an evidence file, allow students to make choices about what they will include. At first, they may not make the "best" choices, but they need practice in judging their own work. As a group, set the criteria for what will be included. You have your priorities, and a district anchor paper might need to be included, but make sure students have at least one choice.

When you introduce the idea of a portfolio to your students, be ready to answer their questions: "Why are we doing this?" "Why now?" "Why can't I keep my work?" Let students know the purpose of a portfolio. Assure them that they will reap benefits from their completed portfolios long after they leave your classroom.

A Sample Process for Introducing Portfolios

You have decided that you are going to try portfolios. What do you do next? Here is one process that you might try. First, you have to answer several questions. Then you need to figure out how you will manage time and space issues. Let's look at the focusing questions first. Spend some time considering the following questions:

- Why do you want to use portfolios?
- What are your goals?
- What purpose will using portfolios fill?
- By using portfolios, what messages do you want to send to students, parents, administrators, and the learning community?

Are you still interested in using portfolios? Yes! Then on to space and time issues.

Objectives After sufficient practice, students should be able to

- collect, organize, and classify personal work into designated categories
- make judgments on quality and appropriateness of a piece of personal work as related to prescribed criteria
- set learning goals and evaluate personal progress toward those goals
- reflect on prior work and use those reflections to guide new learning

Uses Providing evidence of learning. Building a resource database. Building self-esteem. Aiding students in becoming better learners. Allowing for multiple-intelligence activities. Catching the flavor of a classroom experience. Showing a range of skills and understandings.

Directions These directions should be used as a model. How the use of portfolios unfolds in your classroom will depend on many parameters. Re-read the introduction to this section and decide on an exact purpose for portfolios before you set up your classroom for portfolio use. To begin, try using portfolios in only one of your classes.

1. Set up your management system. Duplicate a letter to parents/guardians introducing them to how portfolios will be used in your classroom.

2. Explain to students the what, why, and how of using portfolios. Give students the letter to parents/guardians and ask them to return questions or responses the next day.

3. The next day, discuss any questions from parents or guardians. Have students set up their files. Let students know how and when they will have access to them.

4. From time to time, add items to portfolios.

5. When it is near the end of the unit, semester, or year, have students organize their portfolios, write reflection pieces, and turn in their completed portfolios.

6. Choose whether or not to score portfolios. Use portfolios for their intended purpose.

Follow-Up/Assessment Portfolios will eventually go home, go on to another teacher, be stored as evidence, or even get discarded. Parent comments are worth the effort of sending everything home and hoping that it will all come back again. You might ask students to return just the parent comment pages. This would be the final action with the portfolio.

If you decide to score a portfolio, keep the scoring simple and let your students know exactly how they will be scored. Most of the work that comprises a portfolio has probably already been scored. Don't score students' reflections other than to give credit for completion. In fact, the whole scoring of a portfolio might be based on the completion of required components. Figure 1-9 shows a sample that might be used as a scoring checklist.

Figure 1-9
Sample
Portfolio
Scoring
Checklist

Portfolio Table of Contents

- Student Statement: How Am I Doing?
         Updates 2 (11/93) 3 (1/94) 4 (2/94)   _____
- Focused Free Writing: What Is a Scientist?   _____
- Cause and Effect Analysis   _____
- Open-Ended Essay: Recycling Plastics Paper   _____
- Who Am I?   _____
- First Poem: Haiku   _____
- Cooperative Laboratory Report: My First Analysis   _____
- Graphs: Three Examples of My Work   _____
- Reading Log: One Example of Current Work   _____
- Student's Choice (list category)   _____

Examples The purpose of a *showcase portfolio* is to show student achievement. Students collect and save everything over a period of time. Finally, they take the time to sift through their collection, collate similar works, write reflective statements about their learning, and present their portfolio for evaluation. Handout 1-1: Showcase Portfolio Directions is an example of the directions a teacher might give to help students put together their work. The directions in the handout were used by eighth-grade students in an extensive, end-of-term, self-evaluation. Students had three weeks to complete their organization. They had two days in class to work on the portfolio with teacher help. The rest of the work was completed at home or in the computer lab.

Chronological portfolios are used to show students' abilities to set and accomplish long-term goals. They also make it possible for students to look at a series of products and determine personal growth. They can be used to capture all work regarding one component of learning—for example, laboratory reports, research techniques, or experimental design. Handout 1-2: Initial Goal-Setting Directions and Handout 1-3: Subsequent Goal-Setting Directions were used to guide students' goal setting throughout a school year. After setting goals during the first week of the school year, students revisited goal setting five more times.

In one school, teachers of subsequent grades wanted information about the students who were signing up for their courses. They had specific criteria that they wanted about the students. They also wanted succinct and concise work. In one case, these teachers wanted to know about students' abilities to set and accomplish goals, design inquiries, write in a scientific context, and use scientific equipment. An evaluation portfolio would be a good tool to use for this purpose. Such a portfolio is used to document student achievement in accordance with specific criteria. Handout 1-4: Evaluation and Showcase Portfolio Directions and Handout 1-5: Parent Response Form/Science Portfolio can be used with an evaluation portfolio.

Action Research

How do you know that the activities you use from this book—or any other book—enhance student achievement or cause any other changes in your students' behavior or in your teaching? You could invite an educational researcher into your classroom full time to investigate these issues using current research methods. However, it would be more practical to conduct such research yourself on a simpler, more immediate scale. Action research is one way you can do research in your classroom.

Action research focuses on a question that you would like to investigate. It is teacher-based inquiry and reflection. When you use action research, you are the researcher: you ask the questions, you gather the data, you interpret the results, and you take further action depending on your findings. You share your findings when and where you choose. Action research study teams are formed by teachers who have similar questions about teaching and learning. These teams are often set up so that participating teachers receive district units or, in some cases, post-graduate credit.

How do you go about conducting action research in your classroom? It can be easy and require little time, depending on the questions you decide to explore. All you have to do is pay close attention to what is happening, document events in a concise manner, reflect on what happened, and make changes based on your reflections. You don't have to conduct action research by yourself; form a small group of colleagues to share ideas and results, and to plan further action.

Ideas for action research projects can come from anywhere. You might have attended a conference and brought home a new activity. How does it fit into your current curriculum? Try it, record student reactions, reflect, and make a decision. You might have attended a workshop on a new teaching strategy. Now several of your colleagues are trying to determine the best way to use the new strategy with students. Arrange for someone to videotape you introducing and using the new strategy in your class. Then watch the tapes with other teachers, ask them to provide feedback and comments, and, as a group, make a decision about what worked and what might be improved.

Action research has many useful applications for classroom teachers. It can provide feedback on your teaching practices, and evidence on teaching successes and failures. Many of the first-draft writing activities, such as instant summaries and exit-and-entry slips—strategies that you will learn about later—can provide evidence on teacher effectiveness. Action research can provide students with feedback on their own achievement. Questions used for action research can be consolidated into a student or teacher portfolio. The National Science Education Standards (National Research Council 1996) call for teachers to model how scientists do research. Action research is one way of accomplishing this goal.

Activities to Get Ideas Flowing

2

*T*he activities and strategies in this chapter are designed to help you determine what your students know. The activities should capture your students' imaginations and generate ideas. These activities are all part of the prewriting process. *Prewriting* is any activity done before first-draft writing. The purpose of prewriting is to open a conduit to your students' brains and help ideas flow.

Prewriting activities may be used as topic starters, post-topic reviews, idea-generating sessions, test reviews, or a quick evaluation of your instruction. All of the activities in this chapter can be used as starters for other types of writing.

Some of these strategies may be new to your students. Illustrate examples for them until they feel comfortable with each new technique. Research indicates that it takes twenty-five repetitions before an activity is internalized. (Sassaman 1988) Your students may need a lot of practice.

Prewriting activities are usually stream of consciousness writing activities. If you use them for assessments, they should be scored for effort and completion, rather than style and usage. Encourage students' efforts. At this point in students' writing, criticism may act as a dam to new ideas; encouragement will help build trust that opens doors to more writing.

For increased effectiveness, the first few times you try a new activity, write along with your students—both for your benefit and your students'. Nothing helps students overcome writing fears more than seeing their teacher experiencing the same problems they do. You are their science role model; show them how it is done—fail and succeed along with them.

Strategies in this chapter include:

- Brainstorming
- Clustering
- Questioning
- Focused Free Writing
- Sensory Perceptions
- Guided Imagery

Brainstorming

Brainstorming is a method of generating ideas or topics through group inter-action. Brainstorming can be used with word associations or before clustering. (See next activity.) Scientists use brainstorming to stimulate thinking, and groups of scientists may meet regularly for brainstorming sessions. Brainstorming creates fresh ideas and new viewpoints, some of which can be used for research inquiries. Many inventions are indirect products of brain-storming. Brainstorming as a prewriting activity uncovers many different points of view. Depending on your students' grade level, they may have little infor-mation or a fairly narrow focus on science topics; brainstorming will broaden this focus. There are two rules used for brainstorming: (1) All ideas are wel-comed; and (2) negative criticism is not acceptable—in fact, it defeats the pur-pose of brainstorming.

Objectives After sufficient practice, students should be able to

- Explain the process of brainstorming
- Produce a list of ideas formed as a result of brainstorming an idea or topic
- Write a summary sentence about the ideas produced

Uses Finding out students' prior knowledge on a topic. Reviewing. Generating ideas for research or other projects. Group problem solving. Predicting. Facilitating large group cooperation. Starting a new unit of study.

Directions These directions are for a large group activity. After students have had some practice as a class, they might try brainstorming in smaller groups.

1. Write a topic on the chalkboard. Allow sufficient time for students to think about the topic. Then, ask if the topic needs any clarification.
2. Ask students to think of ideas related to the topic. Then have students share their ideas with the class. Write all ideas on the board.

3. Acknowledge each idea presented.

4. Copy or save the brainstorming ideas for intended use.

Follow-Up/Assessment Score student participation. Use ideas from the brainstorming lists to start other activities. You will find many how-to examples throughout this book. Students will know you value their ideas and input when you use ideas from brainstorming sessions in other activities. Students often will come up with the same ideas you would like to cover in future lessons. This is one way for them to share ownership of what happens in your classroom.

Examples A teacher started a laboratory demonstration on chemical reactions. Part way through the demonstration, she stopped and asked students to brainstorm possible outcomes of the next demonstration. She used this technique often to encourage students to predict possible solutions or to make outright guesses.

At the beginning of a unit on energy resources, a teacher asked students, "What possible alternatives to electricity do we have as a source of energy?" Students produced an array of ideas. These ideas were used later for group research projects.

To start a weather unit that would include a major project, high-school students brainstormed the list of words in Figure 2-1. Other activities in this book will use this list as an example of how easy it is to integrate science disciplines.

---

climate, rain, water cycle, dynamic, erosion, biorhythms, forecasting, seasons, sunlight, migration, heat, currents, animal behavior, ocean, aquifer, pressure, transpiration, respiration, fronts, salinity, density, drought, temperature, careers, data, evidence, clouds, indications, weather as limiting factor, tides, sunspots, pollution, energy, insulation, radiation, cold, agriculture, satellite, satellite lift-off, Doppler radar, air, atmosphere, gas laws, feeding behavior, growth cycles, humidity, models, agriculture, mountains, animal responses, wave motion, thunder, lightning, flooding, tornadoes, hurricanes, typhoons, tsunami, monsoons, wind, energy production, hydroelectric, wind speed, hypothermia, snow, sleet, rain, acid rain, ozone, transportation, micro-climates, salt on the roads, dew point, allergies, static electricity

Figure 2-1
Student
Brainstorming
List—Weather

Clustering

Clustering is a prewriting activity in which writers write a central word on a sheet of paper and jot ideas around it. Circles or other shapes are drawn around the ideas, and lines are drawn between related ideas. Ideas can be expressed in words, phrases, or drawings.

Clustering is widely used as a visual means to express and organize ideas. It is not a linear process, which makes it an excellent divergent-thinking activity. Clustering may be used after brainstorming or as an introductory activity. It is a way of arranging ideas, similar to an outline, but without the rigidity. Clustering is also known as *mapping, treeing, webbing,* or *bubbling.*

Objectives After sufficient practice, students should be able to

- create a cluster of words associated with a central idea.
- organize these words into a logical arrangement for further use
- write a general statement about their cluster

Uses Introducing a new topic. Assessing students' prior knowledge on a topic. Reviewing for an exam. Producing a quick response to an essay question. Generating ideas. Organizing thoughts.

Directions Illustrate this technique as a large group activity until students are comfortable doing it on their own.

1. Write a topic, big idea, issue, or statement on the chalkboard.
2. Draw a circle around what you have written.
3. Ask students to think of related ideas and share them with the class.
4. Write each idea on the board. You may want to group similar ideas in the same area.
5. Draw circles around each idea as it is presented.
6. Discuss how each idea relates to the central topic or other ideas.
7. Continue until no new ideas are presented or time allotted is used.
8. Draw lines connecting related ideas.
9. Have students review all the ideas and write a general statement about the cluster.
10. Save clusters for further use.

Follow-Up/Assessment After students are familiar with this technique, they should be able to create clusters individually. These clusters can be scored for relevance, completion, and effort. You might be surprised at how well this activity works for generating new information.

Examples Students starting an open-ended unit on weather first brainstormed a list of words. (See previous activity.) They were asked to cluster their twenty favorite ideas. One student's cluster is shown in Figure 2-2. Following are the words he chose from the brainstormed list: *climate, forecasting, seasons, sunlight, currents, temperature, careers, data, evidence, energy, insulation, radiation, heat, cold, satellite, Doppler radar, humidity, wind speed, dew point,* and *allergies.*

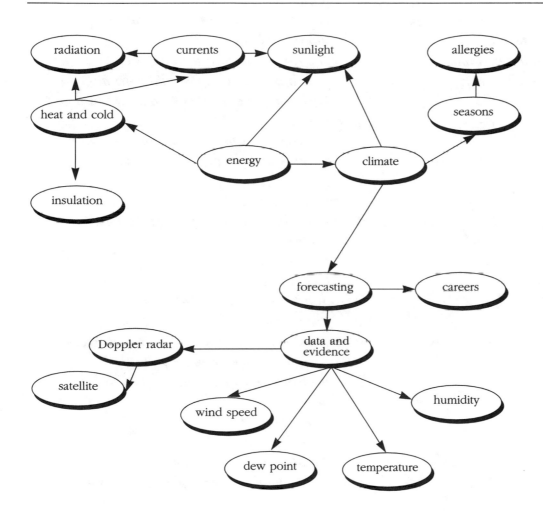

**Figure 2-2
Weather
Cluster**

As an introduction to a unit on light, the following sentence was on the board when students entered the classroom: *What would we do without sunlight?* The class made a cluster and saved it until the end of the unit. For extra credit at the closure of the unit, students wrote poems using any part of the cluster or information they learned during the unit. Figure 2-3 shows two poems that students wrote.

Figure 2-3 Student Poems (using weather cluster words)

"There Is No Light I Can't See Right"

It is so cold I think I'm standing on mold.
There is no food, I couldn't cook.
I heard a noise, I couldn't look.
The atmosphere will freeze.
I've lost my tan.
I'm gonna starve, 'cause I'm out of Spam®.
There is no energy, it's no fun
I'm gonna die of total boredom.
No heat, no light, so what is left
Oh, great
I lost my checkers.

What I Would Do Without the Sunlight

If there were no sunlight,
There would be no animals to bite,
It would be very dark,
And I couldn't play in the park,
If there were no light,
There's one good thing you couldn't fight.
But our wonderful life would end
The rays of the sun wouldn't bend
We would starve in the cold with no food
Then everyone would be bold and in a bad mood.
We would not be able to cook.
There is no heat or we couldn't see in the book
If there was no light it would be boring, no suntanning, no pretty plants to look at, we couldn't
Even swing a bat.
Everything would be unproductive
and it would be very difficult to live.
Even though you say pretty please
With no heat our atmosphere will freeze.
We would have no energy at all
We would not be strong enough to catch a ball.
If we couldn't see
There would be no Vitamin D
So after all of this dread
No sunlight unfortunately means dead.

Questioning

Questioning is a strategy in which students generate questions starting with the words who, what, where, when, which, why, and how. Using these words as starters for questions makes students ask questions requiring more than simple "yes" or "no" answers. *What, why,* and *how* can produce higher-order thinking questions. *Who, where, which,* and *when* can produce descriptive, recall, ordering, and comparative questions. This technique—called the *6WH's*—can be used when students are asked to produce a variety of questions quickly.

Objective After sufficient practice, students should be able to write six to ten questions about a given topic using the 6WH words

Uses Previewing or reviewing a unit. Creating study guides or a question data base. Developing questions for quizzes or research.

Directions This activity can be used for small groups or individuals. When starting a unit, questioning can be used as a variation on brainstorming. The following directions are for using questioning as a unit preview or review with small groups.

1. Identify a topic. Allow time for students to ask questions to clarify the topic.
2. Hand out index cards.
3. Have individual students write six to ten questions about the topic, one question per card. Questions must start with one of the 6WH words. Students may use notes, books, references, or other materials related to the topic to generate their questions. Students write their initials on the question side of the card.
4. If this activity is a preview, collect cards from individuals or small groups. If the activity is a review, have students write the answers on the reverse side of the cards.
5. Have students share their cards with other members of their group. The accuracy of each answer is verified by another group member. The verifier places his or her initials on the answer side of the card.
6. Direct groups to exchange cards and to repeat the verification process. Cards may be initialed again.
7. Collect the cards.
8. To use the cards as a review, have groups exchange sets of cards. Group members quiz each other using the questions and answers on the cards. Students rotate cards to all groups and continue quizzing.
9. You can use the cards as flash cards during a few extra minutes in class. Answers can be used as a game-show quiz for a lively review.

Follow-Up/Assessment Give credit for completion of the assignment. This is a good activity to grade group participation and attention to task. Each card can be scored for appropriate use of content and for the use of higher-order thinking skills.

Examples Students were asked to write questions they have about the top twenty words they chose for their weather cluster. (See previous activities and examples in this chapter.) One student used her cluster to come up with the questions in Figure 2-4.

Figure 2-4 Weather-Related Questions Using the 6WH's

Sample Questions Using the 6WH's, Written by a Student from Her Weather Cluster

HOW do meteorologists forecast weather?

HOW do scientists gather the data to forecast weather?

WHERE is the data for forecasting weather consolidated and shared?

WHERE do weather forecasters get their information?

WHAT kinds of instruments are used to gather weather data?

HOW do these instruments work?

WHAT is the science behind the machines?

WHEN is the best time to insulate your home?

WHAT are the best ways to insulate your home?

WHY do my allergies act up in some houses and not in others?

WHICH is easier to insulate against, heat or cold? WHY?

....

To prepare for a final exam in a high school, integrated, general science class, students divided the semester's content and wrote questions about the content on index cards. Students went through their notes, laboratory work, and projects. With the help of the teacher, they came up with questions at all levels of Bloom's taxonomy. Using these cards, along with other classroom materials, students made individual study packets. They used these packets to review for their exam. What they didn't know until they were almost finished making the packet was that the teacher planned to use the study packets as a major portion of their final score. They took their study packets to the final exam and submitted them with the exam. One packet included the questions in Figure 2-5.

Questions Using the 6WHs, Written by a Student to Prepare for an Exam

Figure 2-5
Using the
6WHs to
Prepare for
an Exam

WHERE would you set up a recreation area based on our land use studies?

WHAT major issues did we study this semester?

WHY are people concerned about water quality? Defend your position.

WHEN is it wise to conserve energy?

WHO was the greatest inventor of times past? Defend your choice and give at least three examples.

HOW did your group solve the erosion problem? Be sure to describe the simple machines the group designed.

Focused Free Writing

Free writing is nonstop writing for a sustained period of time, usually five to fifteen minutes, depending on the sophistication of students or how much practice they have had with this activity. *Focused free writing* is writing on a specific topic. Students write whatever they are thinking, as quickly as possible, without worrying about grammar, punctuation, or style. This type of writing can be an instant tap into your students' imagination and store of knowledge. Focused free writing is often used to start thoughts flowing about an idea. Research indicates that if people write more than four lines, they will continue writing for about five to fifteen minutes. It takes the brain about one and one half minutes to switch into stream of consciousness and engage the hand in free-flowing writing. (Tierney 1989) As a teacher of science, you can use this prewriting activity to open up the creative portion of your students' minds. Using this activity before a discussion will give students time to think about the topic or issue that will follow.

Objectives After sufficient practice, students should be able to

- write continuously on any given topic for five to fifteen minutes
- express ideas clearly

Uses Generating ideas on a new topic. Reviewing quickly before a test. Creating ideas before a discussion. Directing students' mental focus. Finding out students' prior knowledge. Initiating focused thinking on a topic or issue for a following discussion.

Directions This individualized activity is best carried out in a quiet room. Students need uninterrupted time for ideas to develop. Once students begin to write, allow no verbal interruptions. For increased effectiveness, write along with your students the first few times this activity is tried. When introducing this activity for the first time, fifteen minutes may be too long. Start with two to five minutes and build up to fifteen minutes. The first few times you do this activity, you may want to allow students to choose the focus.

1. Write the topic, issue, question, or organizer on the chalkboard.
2. Explain that students will have _____ minutes for nonstop writing.
3. Have students write the topic on their papers.
4. Ask students to begin writing/thinking nonstop for the allotted time. Remember that during part of the nonstop writing time, students' pens or pencils may not be moving. Hopefully, their minds will be!
5. When the allotted time is completed, have students read to themselves what they have written. Then have them write a general statement summarizing their work.
6. Encourage discussion of students' general statements. Ask volunteers to share their general statements with the rest of the class.

Follow-Up/Assessment Score student work for completion and effort only. Encourage all efforts in free writing. Comprehension of a topic might be assessed by the general statement written at the end of the free-writing assignment. Discussions that follow free writing or any other prewriting activity tend to be lively because students have had time to think and write beforehand.

Examples You might begin or end a class with focused free writing. Ask students to recap the day's lesson through free writing. Assessing student comprehension before and after a lesson or class can be done with free-writing admit and exit slips. Samples of these are shown in Figure 2-6. Using these slips, students write a brief response to a question. Admit slips can also be used to direct students' mental focus on your class or on a topic you wish to introduce.

Figure 2-6 **Admit Slip** **and Exit Slip**	ADMIT SLIP Name _____ Date _____ Directions: Tell me in writing a few things about _____ EXIT SLIP Name _____ Date _____ Directions: Summarize today's class lesson in 25 words or more.

One chemistry teacher asked for a free-writing summary at the end of his class. He then spent the opening minutes of the next class clarifying any misinterpretations he found in the summaries.

Students in a general science class studying light were asked to write a story that included their vocabulary words in context. They wrote the story in class as a focused free-writing activity. Later, students rewrote their stories for a bulletin board. The two samples in Figure 2-7 are from this assignment. The vocabulary words are underlined in each sample.

Sample 1

It was the morning of Monday, 17, 198–. I woke up for school. But I thought for a minute and I remembered it was 198– when I went to sleep. So I went to my old house and I saw my reflection of me, but I was 9 years old. My family could not see me, but I could see them. I found when I was always looking in the mirror. I saw my old chandelier break again. "I broke it when I was 9." I remember saving some of the prisms from the broken chandelier. I did experiments with them and found 3 colors red, blue, and green. The 3 primary colors. I put them together and made white light. Then I pulled out my parabolic mirror and saw my deformed image. I found out that was called a virtual image. Then I woke up and I even happy it was just a dream but I saw the old prism on the floor and it made a spectrum on the wall.

(Remember that this is free writing. For a first draft, this student exhibited imagination and understanding of the vocabulary.)

Sample 2

I was walking along a pond near my grandfather's orchard when I saw my reflection in the water. It was like a virtual image, but when I threw a stone in the pond my image became distorted, almost like the rippling waves were an interference but I noticed that each wave was like a parabolic mirror. When my focal point was clear again I saw that the way the water was shown upon by the white light coming from in between the trees onto the water it appeared as a spectrum blending into a white dot of light. I was getting tired of just looking at the water so I decided to experiment with it. So I went to get a stick and tied to turn over a rock, but the rock was further out then I thought. I guess I couldn't reach it because the light was refracting its image. Those little white dots bothered me as much as my brother does, he's all over the place, but not in the place I want him to be. Then I noticed the spectrum around those dots were just the primary colors.

The water was just like a mirror. The colors shown on this "mirror" as if they were made by a prism.

Now I have realized how important this part of the electromagnetic spectrum is....

(This story shows sophistication in its use of metaphors. The writer only remembers at the very end of the story that the teacher is the audience.)

Figure 2-7
Samples of Student Focused Free Writing

Sensory Perceptions

Sensory perceptions are any sights, sounds, smells, touches, or tastes that trigger images in students' minds. The mind recognizes and remembers sensory input. If you can tap into students' storehouse of sensory perceptions, you may have found a shortcut to lengthy explanations. In humans, visual sense is the most highly developed. This activity encourages students to perceive using all their senses.

Objectives After sufficient practice, students should be able to

- describe objects using sensory words
- increase their ability to perceive and describe objects
- respond in writing to sensory experiences

Uses Introducing new units or laboratory activities. Reviewing. Gathering information and describing items. Making observations.

Directions These directions are general. Adapt them to your specific classroom needs. Some students might need a list of sense words to help them describe what they experience. You can use this set of directions in many ways. You might start off with only one sense, perhaps touch or hearing. Then use the same materials again, building up observations to include all the senses. You might also ask students to write their comments about each sense separately. You can state the senses in which they record their sensory perceptions.

1. Prepare a sensory activity. Try to think of activities that allow students to sense differences in texture, shape, temperature, and size.
2. Explain the procedure for the activity. In a large group, brainstorm a list of words that could be used to describe things using different senses. A starting list of visual words might include words related to shape, color, size, texture, and patterns. A list of words related to sound might include metaphors or words related to pitch and loudness.
3. Have students use their five senses to make observations. *NOTE: Students should not taste or eat anything in a classroom unless directed to do so by the teacher.*
4. Have students write descriptions of what they observe.
5. Have students write a general statement summarizing their experience.

Follow-Up/Assessment Students should recognize and describe items using different senses. Have students self-score pre- and post-descriptions. Students should count the number of "sensory" description words and the number of senses used in their descriptions. Award points accordingly.

Examples During a botany unit, different kinds of flowers were placed in individual brown bags where they could not be identified by sight. Students described the odor from each bag and tried to identify the kind of flower in the bag.

During a unit on the sense of smell, students were asked to bring a smell to class. Students filled a cotton ball with their smell and put it into a plastic bag. Each bag was numbered. Students observed each smell and tried to describe it without identifying it. Then students shared their descriptions, and other students tried to identify the smell from the description. Using similes was useful in describing smells.

During a sound unit, students went on a sound hunt. They listened at home, on the way to school, at school, and in class for different sounds. They wrote down descriptions of these sounds without identifying them. In class, students tried to identify sounds from the descriptions. The descriptions of the school bus were the easiest to recognize.

During one summer session, students spent a week studying sensory systems. One group made a tape of sounds as a test for the rest of the class. Sounds included snoring, scissors snipping, brushing teeth, starting a car, and the voices of students from another class. One sound that eluded everyone's ear was the repeated sound of a moving sprinkler.

Activities using the sense of touch may bring back memories of elementary school mystery boxes—the ones where you would put your hand into a box and try to identify the object inside by how it felt. One earth science teacher made a middle-school version of the mystery box. She placed rocks with different textures into a box. Soil samples were used as well, so that students might feel different soil textures.

For a research project, two biology students conducted a taste survey among their peers. They wanted to know if students' taste buds could differentiate between wintergreen and peppermint candies.

Some teachers show students the Search for Solutions (Associates 1980) videotape on patterns. It is useful as an introductory or summary activity in the discussion of patterns.

Guided Imagery

Guided imagery, or visualization, is oral narration that allows students to imagine or role-play. It can be used to help students visualize and internalize difficult science concepts and ideas. Students become directly involved with content. When students create pictures in their minds, they access their prior experiences and create new pictures. This process helps students visualize a situation, so that when you ask the question, "What happens next?" they have the capacity to imagine answers.

Try using the following sequence of activities to help expand students' ability to imagine: Start by asking students to imagine concrete objects. Next, have them recall people and places. In the third step, expose students to readings that include plenty of imagery. Finally, guide students to create their own images from the content you are presenting. This progression allows students to transfer more science information from short-term memory to long-term memory.

The great power of visualization is that it encourages students to recall information beyond what you ask for. You may be able to engage your students with a chosen specific content in a positive manner. Research indicates that it is easier recalling pictures than facts. (Caine and Caine 1991) Give students as many opportunities as possible to create science pictures. Help your students create content pictures and imaginary situations through visualizations.

Objectives After sufficient practice, students should be able to

- listen to a guided visualization
- describe their visualization in writing
- increase their recall of information

Uses Reviewing a unit. Exploring new concepts. Describing a reading assignment. Introducing new ideas. Reducing test anxiety.

Directions This activity should be done with a large group. The group should be quiet and relaxed. Allow no interruptions. Make your first visualizations short, about two to four minutes.

1. If possible, dim the lights.
2. Have students prepare themselves by taking a few slow breaths and by closing their eyes. The first time some students do this activity, they may become tense. Eventually, they should relax.

3. Explain to students that you are going to help them envision a multisensory picture and that they need to be relaxed and quiet so that they can build a picture as you guide them through the journey.

4. Start your visualization. Make it as sensory rich as possible by telling students what to see, hear, smell, feel, and taste.

5. Allow time for several pauses.

6. Include many words that relate to the five senses.

7. As you speak, circulate slowly around the room. Watch for student responsiveness.

8. End the visualization.

9. Have students write freely or draw a picture about their sensory experience or draw a picture.

10. As closure, have a short class discussion on the differences of individual interpretations of the visualization.

Follow-Up/Assessment Have students write short visualizations based on their experiences before and after a unit. Free writing may also be graded for completion. (See Focused Free Writing earlier in this chapter.)

Examples On a tray, display several objects that relate to a science topic. After students have had a chance to look at the objects, remove them. First, discuss where the objects were located relative to each other. Then ask students to describe the objects. Display the tray of objects again and compare them to students' descriptions. This activity makes a great game with objects, such as lab equipment, chemicals, leaves, insects, or animals.

During a unit on fossil fuels, a teacher took her students on an imaginary trip through a coal mine. Afterwards, they wrote about the experience.

In a general science class, students had fun imagining themselves as animals. First, the teacher led them through an imaginary day in the life of an animal. Then they wrote about the life of a specific animal of their choice.

One teacher took his students on a trip through a cell and several microscopic organisms. Another teacher ended the school year by taking students on an imaginary trip through the solar system. Students designed the space ship they traveled in. Some ships came complete with rock groups.

You might also use guided imagery to encourage student success in tests. Have students imagine themselves studying for the upcoming test, taking the test, and doing well.

Ask students to describe the mental pictures they have that help them play video games or memorize plays in sports. You might be surprised at the details they can share with you.

The visualization in Figure 2-8 is a sample used for introducing an appreciation of the different kinds of weather. Adapt this visualization to fit the season and weather of your choice. Remember that students see examples of many kinds of weather, but they experience only local weather. This visualization builds a multisensory weather picture. Violent weather phenomena would also make interesting visualizations.

Figure 2-8 Sample Visualization to Introduce Weather

For this visualization, students should have writing and drawing materials on their desks. Read each statement aloud, allowing various length pauses between statements.

- Sit back in your chair, and take a couple of deep breaths. (Pause)
- Let all the tension out of your body. (Pause)
- Close your eyes and let me lead you through a visualization that will help you experience the weather on a hot, humid, summer day in the Midwest. (Pause)
- It's hot, over ninety degrees. (Pause) (If you've had a day this hot, use it as a reference.)
- It's humid, like a wet sauna or a locker room where all the showers are turned on. (Pause)
- There are no clouds in the pale, blue sky. You forgot a hat. (Pause)
- There is a slight breeze—as if you were sitting by a window fan on low. (Pause)
- There are cornfields all around you. The corn is waist high. (Pause)
- There are birds around, but you can't see them because the sun is in your eyes. (Pause)
- You are walking by yourself on a gravel road between two fields. (Pause)
- What can you smell? (Pause)
- Walk along this road for a while, letting your senses be your guide. (Long, long pause here)
- When you are ready to return to class after your walk, quietly stretch, open your eyes, and begin recording the picture you imagined. (Pause)
- When all students have begun writing or drawing, you can play some music that would set the mood for Midwestern farm country. Allow students to share their drawing with a neighbor. Then have a short class discussion on the different pictures students envisioned.

Writing for Understanding

3

*I*n regards to the history and nature of science, the National Science Education Standards advocate that students understand that science is a human endeavor. Too often the media present science as a process that is completed and cannot be changed. Scientific journals often use a format of reporting results that doesn't reflect the true method of how scientific information is explored and discovered.

Scientific endeavors are full of failures, dead ends, and detours. As such, they reflect how humans learn—by doing, making mistakes, and redoing. The process might be compared to the one a gymnast goes through. She has thousands of hours of practice, healed bruises, and has made mistakes that were repeated many times until her skill level reaches the perfection that we observe. Seldom does the observer get a glimpse of the gymnast's failures and repeated practices. The observer sees only the "finished product." Like the gymnast, a scientific researcher has many trials and errors. These can be found as notes in the researcher's daybook.

Daybooks are used to store thoughts and ideas. A scientific researcher uses a daybook to enter daily notes about laboratory experiments, discussions, reflections, questions, and readings. Daybooks are often used as stepping stones to larger projects. Researchers constantly refer to their daybooks to trace an idea back to its conception, or to follow the growth of an idea into a completed project.

Just like a scientific researcher, your students will use induction and deduction to unravel the flow of ideas when they use daybooks. The information in this chapter will help you provide students with strategies that will help them improve their science comprehension and enable them to become more involved in their own learning.

We do not write in order to be understood. We write to understand.

....

C. Day Lewis

Strategies in this chapter include:

- Science Daybooks
- Note Taking, Note Making
- Reading Logs
- Instant Versions
- Summaries

Science Daybooks

A lot of thinking should go into the work done in a science classroom. Unfortunately, the ideas and thoughts that are generated often get lost because no attempt was made to record them—at least not in one place. Daybooks provide a place where students can store a variety of thoughts. They allow students to trace their own ideas through a class year. Students become more involved with their learning when they write in daybooks. The focused writing that students do in their daybooks should increase their fluency and confidence with science concepts. Ideas and evidence can be stored for future reference.

One key guideline in daybook writing is that students should write frequently and freely. A daybook may contain personal reflections similar to those in a personal journal, but these reflections should relate to a science concept being studied.

Review your students' daybooks often. Frequent review opens another line of communication with your students. Score daybooks for completion and effort, not style and usage; the writing entered into a daybook is prewriting and first-draft writing. Remember that the red pen can be an instant stop sign to creative ideas. Instead, use a variety of rubber stamps, short comments, or your initials and the date to note completion and to encourage creative ideas. If you are in a self-contained class or have a small number of students, start a dialogue, as explained in the Note Taking, Note Making section of this chapter.

The physical makeup of the daybook is up to you. Keep it simple. The daybook should be used all year. You might have students divide it into sections, in which case a loose-leaf binder would be appropriate. A composition book might be used, resembling a researcher's logbook in which each day's work is entered and dated. Even in these days of electronic note keeping, scientific researchers are required to handwrite their daily notes.

Objectives After sufficient practice, students should be able to

- increase understanding of science concepts, rather than memorizing facts
- use writing as a process for discovery
- improve their personal management of ideas and information
- recognize a connection between thinking and writing
- write freely and more often

Uses Prewriting activities. Starting lessons. Bridging the gap between a student's former activity and your class. Substitutions for "pop" quizzes. Introducing new subjects. Summarizing. Solving problems. Recording discussion notes and laboratory data. Asking questions.

Tips for Successful Daybooks

- Have students organize their daybooks to your specifications.
- You may wish to have students keep their daybooks in your classroom so that they are always available.
- Schedule plenty of writing time.
- Encourage students to re-read and summarize earlier entries.
- Collect and read student daybooks often.
- Encourage students to write questions in their daybooks. They might even set up a special section for questions.

Follow-Up/Assessment A daybook consists of first-draft free writing. It is a place for students to record their ideas. Don't score ideas, nor should you grade style and usage. Explain to students from the beginning that daybooks will be scored for effort and completion.

One fun way of scoring daybooks is to use an audiocassette recorder. Ask each student to bring you a tape. As you score the daybook, make comments into the recorder; then return the tape and daybook. Your students become responsible for listening and making changes to their work. One high school teacher used this technique with his high school biology classes. His students showed more interest in their daybooks and class activities after listening to the teacher's taped remarks. This is a good technique to use with English as a Second Language (ESL) and Limited English Proficiency (LEP) students.

Examples Ideas that make interesting daybook entries might include

- any of the Chapter 2 prewriting activities
- first drafts of short reports

- focused free-writing activities at the start or end of class
- interrupt a discussion with a few minutes of writing
- class summaries
- literary techniques that can be adapted to your instruction. (See Chapters 6 and 7 for ideas. You might also ask an English teacher.

Following are some possible topics:

- Begin an entry with "What if _____
- Begin an entry with "I wonder _____
- Describe a field trip you would like to take.
- What do you know now about _____ that you didn't know before?
- What would you like to know about _____ ?
- What questions do you have about _____ ?
- Explain to (appropriate audience) how to (fill in the topic).
- Describe how to (current process that needs explaining).

Figure 3-1 shows an example of daybook assignments for one week of school. Comments within parentheses refer to activities presented in this book.

**Figure 3-1
Sample
Daybook
Assignments**

Day 1: *At the start of class:* "Today you will begin your science daybook." (Describe daybooks.) "Please begin by writing 'Table of Contents' on the first page. Skip four or five pages into your book, and write the date in the upper outside corner of the page. For your first entry, answer the questions on the board." (See Focused Free Writing, Chapter 2.)

The questions that you have written on the board are:
1. What is science?
2. Describe a scientist.
3. If you were in a place where scientists were working, what would be going on? Describe what you would see, hear, or observe.

At the end of class: "Please use these last few minutes to summarize what happened in class today. You should have at least two main ideas." (See Summaries, later in this chapter, and Focused Free Writing, Chapter 2.)

Day 2: "Today we will start the note-taking, note-making part of your daybook." (See Note Taking, Note Making in this chapter.) Prepare a ten-minute discussion of a current issue or concept.

Following the discussion: "Now please write a general statement about the main point of our discussion." (See Summaries, later in this chapter.)

Day 3: *At the start of class:* "Open your daybooks to a new page, write the date in the upper outside corner, and respond to the question on the board." (See Focused Free Writing, Chapter 2.) To spark student interest in a new chapter, the question on the overhead might be: "Pretend you are one of the following people: an environmentalist, a chemist, a factory inspector, or a medical doctor. Make a list of how energy might be used in your field." (See Brainstorming, Chapter 2.)

At the end of class: "For homework tonight, choose a topic from the brainstorming list we made at the start of class and write some 6WH questions on the topic. Bring your questions to class tomorrow." (See Questioning, Chapter 2.)

Day 4: Group students according to the occupations they choose on Day 3. "Share your homework questions with your group. As a group, choose one major and three minor questions about your topic that you think you can investigate. Make a cluster of possible experiments and research you would like to explore during the next week. Record these possibilities in your daybook." (See Clustering, Chapter 2.)

Day 5: Have groups of students define the parameters of the major experiments they would like to do in class. Then direct them to write procedures, make a list of materials, and design the data tables they will need to record their observations in their daybooks.

Note Taking, Note Making

Note taking paired with note making is a method of taking notes. While note taking is an objective summary, note making is subjective, prompting students to inquire into what they are studying. Note making also allows for students' prior knowledge as they listen to a discussion. Students stop looking for "right" answers and start interacting with content. The questions your students ask in their note making will give you instant feedback about their understanding of your presentation. You might choose to address their questions during future discussions.

Note making will encourage students to be active listeners. When students discover that you are interested in their queries, they will probably respond with increased enthusiasm. Students' questions can also be used as topics for research projects. Note taking and note making may be separate writing assignments or an integral part of student daybooks.

Objectives After sufficient practice, students should be able to

- respond to relevant questions and problems
- write questions regarding class sessions, pose hypotheses, and identify apparent inconsistencies, relationships, and original principles

Uses Providing an alternative form of note taking. Providing instant reaction to discussions. Providing an outlet for shy students. Helping students absorb background material. Involving students with content. Assessing student understanding.

Directions Note Taking, Note Making is used during teacher lectures or class discussions. The sample in Figure 3-2 shows the results of a note-taking, note-making activity.

1. Have students divide a sheet of note paper in half vertically.
2. Tell them to label the left column Note Taking and the right column Note Making.
3. As the class discusses a topic, have students write notes in the Note Taking column.
4. During pauses, tell students to write their reactions on the Note Making side of their papers. These reactions might be questions, clarifying statements, hypotheses, ideas for further areas to explore, or other comments on adequacy and consistency of evidence.
5. Take time to discuss students' comments and correct any naïve conceptions.
6. At the end of the discussion, have students write a general statement or summary statement at the end of their notes.

Follow-Up/Assessment Note-taking, note-making exercises should increase content internalization by students. Students need practice in thinking about classroom discussions, in writing down what they hear, and in identifying what is important. Sometimes things are said so quickly during class that students miss out on key ideas. The first few times students do this activity, it might help if you collect and comment on students' notes. When you show an interest, students will generally take more time and effort in their work. Students may become better listeners during discussions when you ask why an important main idea did not make it into their notes. Students might realize that their notes have value, and they may attempt to take clearer and more concise notes. If you score note-taking, note-making writing, remember that it is first-draft free writing. You might want to score it for completion only.

Example The notes in Figure 3-2 were taken during an introduction to general energy concepts at the middle school level.

Note Taking	Note Making
Energy comes in many forms. Examples from us were: heat, sunlight, machines, electricity, food, oil.	*What is energy? She didn't define it so I could write it down. I wonder what will be on our test?*
Energy is needed to make things go, run, or happen. Examples came from us, some said, cars use gas, computer uses electricity, CD uses a laser, and electricity, I use food, weather uses the earth and sunlight.	*I have a snowmobile, and it uses a lot of gas. My bicycle uses my power. We have an electric lawn mower. The grass uses sunlight.*
Energy transforms into other energy. Examples from us were, photosynthesis, wind to electricity, water to electricity, electricity to heat, oil to heat, oil to heat to mechanical, electricity to mechanical, chemical to electrical.	*I'm glad we made a list of energy types for this part. I was thinking of how I use batteries for my CD, and plug in the toaster.*
Changes in energy cause all sorts of problems in our lives.	*We have to write about this for homework. One thing is pollution, I know, and heat is a second.*
Summary Statement: Energy is all around, it does different things, it changes. I use lots of energy every day.	

Figure 3-2
Student Note-Taking, Note-Making Sample

Reading Logs

Reading logs are a spin-off of the Note Taking, Note Making activity. Students summarize reading content on one part of a page, then write their comments or questions on the other side of the page. Such logs encourage student interaction with text, and challenge students to read, write, and think about the reading material. As a result, students practice many thinking processes, including outlining, summarizing, recognizing meaning, identifying key elements, evaluating reading material, analyzing, and synthesizing. Reading logs are especially good for ESL students; encourage them to write their comments in their native language, to look up the definition of new words, and to write the new words in the comments column.

After reading and writing, students should discuss their reading. Both small groups and large groups work well. This sharing gives students the opportunity to find out what other students learned from their reading. Discussion will help students improve their higher-order thinking processes as they listen and share their own ideas. Reading logs may be separate writing assignments or an integral part of student daybooks.

Objectives After sufficient practice, students should be able to

- increase reading comprehension in a content area
- increase critical faculties in assessing what they read

Uses Analyzing any written work. Building student confidence with written material.

Directions Before students try this activity on their own, you might lead the class through a sample paragraph or two. Choose an interesting science paragraph from a newspaper or science journal and duplicate it for the class. Read it aloud as students follow along.

1. Have students divide a sheet of paper in half vertically.
2. Tell them to title the left column What's Happening and the right column My Comments or Questions.
3. Allow time for students to read the text.
4. Tell students to stop after every paragraph or other logical break in the reading. Then, in the left column of their papers, have them summarize in their own words what the text says. In the right column, have them write any questions or comments.
5. After they have listened to the entire reading, have students write a general statement summarizing the reading assignment.

Follow-Up/Assessment The first few times students do this activity, their comments may be shallow or trite. As the year progresses, students may surprise you by becoming excellent reading critics. With questions already written in their reading logs, students should have more to say during classroom discussions. Logs should be scored for effort and completion. Any change in student achievement and attitude is an added reward.

Example The format for a reading log is similar to that in Note Taking, Note Making.

An example is shown in Figure 3-3. The reading log was written while the student read an article on new brain research.

Figure 3-3
Sample Reading
Log

What's Happening	My Comments or Questions
Some scientists think the brain is like a muscle—the harder you use it, the more it grows.	*How hard do you have to use it for it to be fit?* *Is there a brain exercise program?*
Individuals have some control over how healthy and alert their brains remain as they get older.	*What do I have to do to keep my brain alert?*
Doing mental exercises may cause brain cells to branch and form new connections.	*How do more connections help me think better?*
General Statement: It's important to use my brain so I stay alert as I get older.	

Instant Versions

Instant versions are focused free writing. Students write freely on a given topic for a specified amount of time. Students might use any number of writing styles, including narrative, descriptive, fictional, factual, a letter to (specify an audience), journal entry, problem-solving, analysis, point-of-view, synthesis, evaluation, persuasion, argument, or debate.

Objective After sufficient practice, students should be able to write an instant version on a given subject using a variety of writing styles.

Uses Prewriting activities. Starting lessons. Pre-discussion starter focusing on current questions. First draft writing check for understanding.

Directions This is an individual activity. Decide beforehand how brief the writing will be. Perhaps you want only a sentence, a general statement, or a paragraph essay. Give students a specified amount of time to complete the writing assignment. As students become more proficient in this activity, you will want to lessen the writing time. Discuss beforehand with students how they will be scored on this activity.

1. Explain to students that they will be given a limited amount of time to complete the writing assignment.
2. Distribute a copy of the writing directions, or write them on the board.
3. Read the writing directions with the class, and discuss any questions.
4. Set the timer and let your students write.
5. At the end of the allotted time, collect student writing.

Examples In one class, where students were getting ready for a class debate on energy, the teacher posted the following writing directions: In several paragraphs, describe at least three different points of view on the energy issue you have just finished researching.

In another class, a teacher wanted to find out if students needed to investigate weather instruments further. She gave students this assignment: In one paragraph, describe the instruments you would use to make a weather forecast.

Summaries

Summaries are one variation of an instant version. Summaries benefit both teacher and student. As students recap a period or a week of activities, they must focus on what happened. Knowing that a summary must be written at the end of class should motivate students to become better listeners.
You can use summaries as instant feedback on new teaching practices. You can also assess student understanding of a day's work. Depending on the feedback, you can adjust the next day's lesson. Ask students to write a summary at the conclusion of any activity.

Objectives After sufficient practice, students should be able to

• summarize a class activity or reading assignment
• produce feedback on a lesson

Uses Review instant feedback for teacher. Assess student understanding.

Directions At the conclusion of any activity, ask students to write a summary. This can be done in class or as a homework assignment. Students can work in groups or as individuals.

Examples At the end of a class period, one teacher asked students to write on a slip of paper what happened in class. Students submitted these on their way out the classroom door. The teacher called these summaries Exit Slips (Chapter 2, Figure 2-6) and used them to assess comprehension.

An earth science teacher had his students write outlines of the class activities eight times each quarter. His students never knew when he would assign the summaries. He noticed that when all the outlines were compared at the end of the quarter, students had improved their ability to summarize the activities and to organize thoughts, and showed improved comprehension of science concepts. While the teacher didn't grade the summaries for style and usage, he did comment on students' grammar. This encouraged students to focus on major points of good writing.

Science Process and Higher-Level Thinking Skills

4

*T*his chapter is designed to help students think about and recognize the mental processes they use to answer a question. At first, your students may have trouble; metacognition (thinking about thinking) is hard for most people. But once students gain this experience, they have another problem-solving tool. Some of the techniques in this chapter may be new to your students. Use illustrative examples for them until they feel comfortable with each new technique. After the initial introduction, each technique can be incorporated into your regular lessons throughout the year.

The activities in this chapter involve students in a variety of thinking processes. Cognitive processes explored include identification, recognition, classification, integrating material, evaluating for consistency, making decisions, and justification. Students will develop oral and written communication skills in science by analyzing reading material and writing paragraphs using different thinking processes.

Most of the activities in this chapter can be divided into short segments that can be worked easily into your regular schedule. Assign parts as homework and/or as end-of-class summaries. All of the strategies in this chapter would make excellent cross-age activities.

Strategies in this chapter include:

- Classification
- Fact and Opinion
- Cause and Effect
- Comparison
- Gentle Persuasion
- Approaches to Problem Solving

Classification

Classification is the process of grouping things that share common, recognizable characteristics. Grouping is a fundamental thinking process used in organizing data. Although students have been engaged in classification activities since they first sorted the blue balloons from the red, they need experience with classifying data. As students advance in their science courses, they need to advance from classifying physical objects to working with words and ideas.

Objectives After sufficient practice, students should be able to

- list the steps involved in classifying information
- organize material by common characteristics
- set up different classification systems
- classify information in a variety of science content areas

Uses Developing scientific generalizations. Organizing and describing scientific information or data.

Directions These directions focus on the thought processes involved in identifying the process of classification. After this introductory lesson, you can incorporate classification experiences into your regular curriculum using some of the other example activities. Besides using the handout, you can also use materials that students can classify by grouping them on their desks. This is both a large and small group activity.

1. Write *classify* on the chalkboard. You might also start this activity by giving your students a variety of different materials and ask them to organize it all into some sort of order. After they do that, then ask them to define what they did.
2. Ask students for definitions and write their responses on the board. You may receive answers such as "categorize," "group," "sort," or "put things together."
3. Decide on a working definition. Have students write the working definition in their daybooks.
4. Distribute Handout 4-1: Classification of Animals. Or create your own handout for the topic you are currently exploring.
5. Have students group the organisms. Students might group organisms in several different ways.
6. Have students give each group a title that reflects the characteristics by which group members were chosen.

7. Ask students to think about how they divided the organisms into groups and to write down the mental steps they followed to form their classifications. What thinking processes did they go through to group the animals?
8. Discuss the mental processes students wrote down, and list them on the chalkboard.
9. Arrange these thought processes in a logical order. Discuss.
10. Have students write a step-by-step method for classifying.
11. As homework, ask students to make their own classification sheet, using the newly developed procedures they have written. The classification sheet must contain pictures or written descriptions of fifteen items from the student's room or personal space at home. Students should classify these items into groups and give each group a title. They should identify the characteristic they used to put each item into a group. Finally, students should go back through the procedure they wrote and identify the sequence they used in making up their classification sheet. These classification sheets can be used for class practice (with your edits, of course), for scoring, or for a bulletin board.

Follow-Up/Assessment Analyzing their own thinking could be a new experience for students. Some will be able to tell you exactly what they did; for others it may be difficult. Discussing the mental procedures is very important the first time you present this skill and during subsequent reviews and practices. You can score finished work and daybook entries. Other activities in this book, such as the word mapping and many of the prewriting activities, generate lists of items that can be pictured for classifications. This type of activity will help ESL and LEP students build vocabulary. It will help all students visualize many patterns and ways of connecting science ideas, topics, and facts.

Students will need plenty of practice at the concrete and abstract levels before they internalize this new skill. Most of the activities in this chapter will strengthen your students' skills at classifying.

Examples You may wish to review the definition and method of classifying regularly, until all students feel confident with the process. Use questions such as the following to help students:

- How do you divide one group of items into three separate subgroups?
- Write a general procedure for classifying anything.
- Define *classification*.

After students have completed Handout 4-1, you might ask them to bring in pictures related to a topic and then use them for classification activities.

Fact and Opinion

Two important skills in the process of decision making are (1) recognizing what a fact is and how it is determined to be a fact, and (2) understanding the difference between fact and opinion. In science, a *fact* is a statement considered true beyond any reasonable doubt because it is based on repeatable evidence. An *opinion* may, or may not, reflect sound scientific fact. Human value systems constantly act to influence how we form opinions. We often do not recognize or consciously identify and justify these values with how we form opinions. Instead of simply accepting another person's opinion, students should be skeptical; skepticism is a characteristic of scientists. Students often are swayed by others who don't have all the facts to back up their claims or arguments. Gathering facts and using them to support an opinion is a form of induction; a number of facts are used to form a conclusion. Students should be fact finders. Challenge students to support main ideas with facts rather than opinions.

Objectives After sufficient practice, students should be able to

- define fact and opinion
- differentiate between facts and opinions using science content material
- write a personal point of view based on a collection of facts rather than opinions

Uses Writing short essays. Analyzing media presentations. Identifying the bias of reading materials and advertisements. Gathering evidence for other writing activities.

Directions This set of directions focuses on the thought processes involved in differentiating fact from opinion. It is a large group and individual activity. After this introductory lesson, you can incorporate fact-and-opinion experiences into your regular curriculum using some of the other example activities. This activity is designed to use with Handout 4-2: Fact or Opinion? but you can create your own list of current content facts and opinions to use with this activity.

1. Write *fact* and *opinion* on the chalkboard.
2. Ask students to define each word. Write their responses on the board.
3. Decide on a working definition for each term and have students copy these into their daybooks.

4. Read the list of facts and opinions from Handout 4-2. As you read each statement, ask students to identify it as either a fact or an opinion. You can have students write their responses on a copy of the handout. Or you might read aloud the statements and have students answer orally.
5. Discuss student responses. Correct any naïve conceptions.
6. Ask students to explain what they were thinking of when they distinguished facts from opinions. What were their mental processes?
7. Write their responses on the board and ask students to prioritize them into their daybooks.
8. For homework, ask students to write three to five facts and an opinion about a current topic.
9. Discuss the homework and correct any naïve conceptions.

Follow-Up/Assessment Score papers on students' ability to distinguish between fact and opinion. Practice with this activity should enhance students' ability to evaluate and argue in writing and oral presentations. By providing practice in using the higher-order thinking skill of analysis, this activity will help students discern between facts and opinions of various issues. To help students with this task, each time they give a fact or opinion, ask them "How do you know?"

Examples Plan many short fact-gathering assignments as regular unit projects. Students may create lists of facts on a number of topics. With many real-life issues, being able to identify the facts is crucial. Many issues are so cloaked in opinions that decision making becomes more difficult. Here is an activity you can repeat several times, changing the facts to keep it interesting.

1. Review the definitions for *fact* and *opinion* and the processes used in distinguishing the two.
2. Assign a reading passage to small groups of students. Or you can have students choose news articles.
3. Have individual students write freely a list of facts and opinions they uncovered in their reading.
4. Ask students to share the information with their groups.
5. Have groups compile lists of facts and opinions. If no opinion is stated, students could write an appropriate opinion to accompany the article.
6. Discuss each group's work.
7. Collect work for scoring.

Answer key to Handout 4-2: Fact or Opinion? 1-O, 2-F, 3-F, 4-O, 5-F, 6-F, 7-F, 8-F, 9-O, 10-F, 11-O, 12-F, 13-O, 14-F, 15-F, 16-F, 17-F, 18-O

Cause and Effect

In addition to learning to distinguish fact from opinion and identifying the values implicit in various opinions, it is important for students to be able to identify cause and effect in forming a rational opinion. Often, it is difficult to distinguish between causes and effects because they can occur in chains of events. An effect may have several different causes, and a cause may lead to several effects. This activity will give students experience in identifying causes and effects.

Objectives After sufficient practice, students should be able to

- distinguish between causes and effects
- recognize cause-and-effect relationships
- examine scientific issues for cause-and-effect relationships

Uses Writing essays. Evaluating reading assignments. Writing reports.

Directions After this large group introductory lesson, you can incorporate cause-and-effect experiences into your regular curriculum. This set of directions focuses on the thought processes involved in identifying causes and effects. Use the directions with Handout 4-3: Cause and Effect. Or create your own handout that is specific to your current content.

1. Write the words *cause* and *effect* on the chalkboard.
2. Ask students to define both words. Write the definitions on the board.
3. Discuss student definitions until a working definition for each word is agreed upon. Have students write these definitions in their daybooks.
4. Distribute the Handout 4-3. Have students read Part 1, which identifies causes and effects. Discuss.
5. Have students complete Part 2 of Handout 4-3 by matching causes and effects. Discuss responses.
6. Have students read the statements in Part 3 and write a cause or effect for each sentence. Discuss.
7. Ask small groups of students to brainstorm three different cause-and-effect relationships, one from their own lives, one that reflects school activities, and one that is relevant to science class.
8. Help students identify their thought processes by using questions such as the following: "How did you distinguish between cause and effect? Describe the process you went through." Students should write their responses in their daybooks.

9. Ask students to share their processes. Write these on the board, then organize them into a logical order. Students should copy this procedure into their daybooks.

10. Review the definitions of cause and effect.

Follow-Up/Assessment   Score completed work. Award points for the section in their daybook's direction #8. Answers for Handout 4-3 will vary. (See Figure 4-1 for sample answers for Part 3.) If students can support their responses logically and to your satisfaction, accept their answers. When students can recognize cause-and-effect relationships, they will be able to predict outcomes of experiments and other classroom activities with more success. In this way, you lead students away from searching for "the one right" answer toward thinking of several possibilities. This is the essence of scientific inquiry.

You can use a variation of this activity to test how well students write. Use student-written directions for a science activity. Have students write each step of the procedure on an individual index card. Mix up the steps of the procedure. Have students reassemble the work, justifying the order of the cause-and-effect relationships they find.

Examples   When studying the environment, some cause-and-effect relationships are fairly easy to identify. Ozone depletion causes an increase in the amount of ultraviolet light reaching the earth. Burning of organic material (wood, oil, coal) causes air pollution.

Other cause-and-effect relationships are not as easy to identify. Acid rain can damage trees and changes the pH of lakes. But acid rain is an effect of sulfur and nitrogen compounds being released into the air. Therefore, sulfur and nitrogen compounds damage trees and change the pH of lakes. But many sulfur and nitrogen compounds are released in the air as byproducts of the production of manufactured goods. Therefore, manufacturing factories can be considered the cause of damage to trees and changes in pH of lakes. But who runs the factories?

As homework assignments, you can ask students first to identify cause-and-effect relationships from a list, such as in Part 3 of Handout 4-3. (See Figure 4-1 on page 60 for examples.) When students become proficient with these lists, start them writing their own cause-and-effect examples.

Figure 4-1
Sample Causes and Effects for Statements

Cause	Statement	Effect
The heat was left on.	The flask boiled over.	All the liquid in the flask turned to steam.
It was not yet passing time.	The bell didn't ring.	The clapper was gone.
All the cross traffic stopped.	The light turned green.	I drove through the intersection.
There was an oil spill.	Wildlife dies from eating oil.	The food chain was disrupted.
The weather was perfect for plants this summer.	The crops were abundant this year.	It took a lot of work to harvest the crops.
There was an ice storm last night.	The road was icy.	We drove with caution.
The modem was unplugged.	The modem would not connect.	I could not get online.

Below are some short activities you can use to reinforce the identification of cause-and effect-relationships.

In the Laboratory With a laboratory partner, have students review the laboratory work they just completed and identify cause-and-effect relationships in their report. Ask students to list the relationships on a separate sheet of paper, or identify each with different colored markers.

After a Reading Assignment Have students analyze their reading assignment for cause-and-effect relationships. Ask students to write these in their daybooks. (This would be a good time to use the Reading Logs activity from Chapter 3.) Discuss. Ask students to write how they identified the causes and the effects. What thinking processes did they go through to identify the differences? Clarify any naïve conceptions.

Comparison

Comparison questions may be the most-asked type of science essay question. Before students can answer comparison questions, they have to be able to identify attributes of the two items that are being compared. Then students have to list these attributes in a logical order to answer the question.

Objectives After sufficient practice, students should be able to

- define the word *comparison*
- use list making as a prewriting technique in answering a comparison question.
- write a comparison using focused free writing

Uses Discussing sections of laboratory reports. Answering open-ended essay questions.

Directions This set of directions focuses on the thought processes involved in writing comparisons. This is a large group activity. After this introductory lesson, you can integrate comparison questions into your regular curriculum using some of the other example activities.

1. Write the word *comparison* on the chalkboard.
2. Ask students for definitions. Write these definitions on the board. Discuss.
3. Agree on a working definition and have students write it in their daybook.
4. Distribute Handout 4-4: Comparisons. Have students read Part 1, which deals with energy comparisons. Then ask students to add characteristics to the lists already provided for them on the handout. Discuss.
5. Repeat with Part 2, which compares a hurricane and a tornado.
6. Have students write some comparison ideas of their own and create a list of attributes for each subject. You can brainstorm a list of comparison ideas in class and have students do the attributes as homework. Discuss.
7. Have students brainstorm the procedures they might go through to answer a comparison question. What are they thinking about? Write their responses on the board.
8. Order the responses. Have students write the procedure into their daybooks.

Follow-Up/Assessment Comparisons are almost taken for granted in educational writing. In a science context, they can be the focus of many experiments and investigations. While comparisons are not usually considered a higher-order thinking process, they are the groundwork for other types of thinking and writing activities. Cause-and-effect relationships and fact-and-opinion state-ments are based on one-to-one comparisons. For a class assessment, have groups of students share the procedures they wrote for answering a compari-son question. Make a class chart of their best work to use as a performance guide in subsequent assessments.

Examples You can have less experienced students start by creating lists of characteristics of two related items. Ask students to pair the characteristics by attributes and then write an essay describing the two topics. For some stu-dents, just getting all of the similarities and differences separated will be an accomplishment.

Prewriting lists are an important part of the mental process in writing compar-isons. Students set up parallel structures during this time. You can almost hear the wheels spinning as students think through the characteristics of the sub-jects they are comparing.

The following activity can be used anytime after the introductory activity. It can be used before a semester test or final exam. This can be used as a large group activity homework.

1. Review the scientific definition of comparison.
2. Write a comparison question on the chalkboard. You can use Handout 4-5: Comparison Question.
3. Have students copy the question onto their handout and analyze the ques-tion by listing the similarities and differences between the two subjects. Discuss.
4. Ask students to free-write an answer to the question.
5. Have students share their answers in small groups and make corrections before a class discussion.
6. Discuss and correct any naïve conceptions.
7. Distribute your own examples for homework. You can use the ideas from the list below.
8. Discuss homework. Collect and score student comparisons.
9. Try to use at least one of your student-created questions on the exam.

Here are some sample comparison ideas.

Life sciences List several characteristics of the following: vertebrates and invertebrates; mammals, reptiles, amphibians, birds, and fish; cell structures and life systems; food webs and food chains; biomes and habitats

Physical sciences List several characteristics of the following: gases, liquids, and solids; sound and light; renewable and nonrenewable energy sources; levers and screws; acids and bases; K.E. and P.E.

Earth sciences List several characteristics of the following: plate boundaries and the mid–ocean ridge; tornadoes and hurricanes; fresh and salt water; earth and any other planet of the solar system; moons and comets; rocks and minerals

Weather List the characteristics of several different instruments scientists use to gather information about weather conditions. Pair each instrument with the weather phenomenon it was/is designed to measure.

Energy List several characteristics of different forms of energy. List several characteristics of ways humans use energy. Pair as many forms and uses as possible.

Gentle Persuasion

Persuasion is the art of getting others to agree with you. It requires students to know their subject and their audience. When students write persuasive essays, the audience shifts from the teacher to the person or group they wish to persuade. The objective becomes to convince the writer's peers, parents, or someone else to agree with the writer on an issue. Persuasion is generally more effective than arguing in getting a point of view across.

Objectives After sufficient practice, students should be able to

- define the word *persuasion*
- organize information in convincing sequences
- write a persuasive essay

Uses Writing essays. Preparing special projects. Writing letters. Presenting an oral point of view.

Directions This activity may be done over several days or weeks, depending on the issue and time allotted. This activity is for small groups and/or individuals. Be sure to ask students in advance to save all rough drafts so progress can be assessed as the exercise progresses.

1. Write *persuasion* on the chalkboard.
2. Have students create a cluster for *persuasion* in their daybooks. (See Chapter 2.)

3. Discuss their ideas. Contrast *persuasion* to argument.

4. Discuss the different meanings of *persuasion*. Create a working definition. Have students write the definition in their daybooks.

5. Discuss *audience*. Explain that when a person wants someone to agree with him or her, the person needs to know almost as much about the audience as about the topic.

6. Choose an issue. You can choose something that is important to your students for this first assignment. It might be a local science-related issue.

7. Have small groups of students brainstorm ideas relating to the issue. They should write these ideas in their daybooks.

8. Ask students to research the issue. They will find that making comparisons, identifying cause-and-effect relationships, and activities separating fact from opinion will help them with their research. Students should list any evidence they can find that supports their point of view, as well as information from other points of view that they can refute.

9. For homework, have students write a rough draft of an essay that attempts to persuade others to agree with them about the issue.

10. Distribute Handout 4-6: Persuasion Review. Have students share their rough drafts in small groups. As each student reads his or her draft, other members of the group should evaluate it using the handout.

11. Have students rewrite their rough drafts using the reviews as guidelines.

12. Allow time for volunteers to read their essays in class. Take an unofficial vote to see if your students would agree with each person's persuasive essay.

13. Collect each student's prewriting, rough draft, response reviews, and revised draft for scoring.

Follow Up/Assessment This activity can be extended outside the classroom. If students are writing thoughtful essays, they could be sent to the intended audience. Your students may feel strongly enough about some issue to send a letter. If students receive a response from their intended audience, they could share the response with the class. Students need to know that their opinions are heard and acknowledged.

This assignment can be easily scored using a checklist. Figure 4-2 shows a sample checklist you can modify for your own use. Add items that are important to you. Set up the point scale to your specifications and for the skills you want to emphasize.

Item	Points (100 Total)	Completed (Check when completed)
Made a cluster	10	
Wrote definition in daybook	5	
Participated in small group brainstorming	10	
Wrote and turned in rough draft of essay	25	
Completed peer reviews	25	
Turned in prewriting	5	
Turned In response reviews for paper	5	
Turned in revised draft	15	

Figure 4-2
Sample Persuasion Checklist (used to document class-room participation)

Examples For lively entertainment and an excellent example of persuasion, invite members of a debate team to present a short demonstration debate for your classes. They should start anywhere in their debate and argue for five minutes pro-topic and then switch sides and argue against the topic for five more minutes. Afterwards, your students could free-write their reaction and responses in their daybooks. It might amaze your students to see how the debaters argue both sides of an issue with equal expertise.

Gentle Persuasion

My students wanted to start a recycling program at our middle school. We didn't know how, if, or what we needed to make it happen. But we knew we would have to persuade many people to buy into our idea. Here is how we went about it.

First, we brainstormed what we knew about recycling. We came up with items such as the following: saves natural resources, (does it?), government, big bother, costs more money than buying new materials, who does it help anyway, messy, uses lots of water, industrial wastes, things we might recycle, aluminum, metals, glass, plastic, organic stuff, clothes, newspapers, tires, teaches us not to waste our resources, help people less fortunate, attitude of helping others or ourselves, keeps tons of stuff out of the landfills, politics, does it really help, earn money for school projects, makes money from waste.

From our brainstorming session, several groups of students went to research different aspects of recycling. I went to the principal for approval of our idea. It was readily given on the condition that the project be self-sustaining and not add any costs to our already overloaded budget. After hearing classroom reports on the different aspects of recycling that each group researched, we went on a field trip to our local recycling and waste management center. There we talked with the people who took care of these things.

Most of my students decided that they wanted to move on to other issues, but a group of students wanted to stay after school to plan and make this exercise become a reality. Our Science Club had a new purpose. We made a list of priorities that included why we wanted to recycle, what we thought we could recycle, how we would recycle, how we would convince the rest of the school that it would be a good idea to recycle, and what each of us could do to make recycling happen at our school. We decided we could tackle the biggest and cleanest of our school's wastes— office paper. Our local recycling center provided us with collection bins—enough for each classroom and extras—and a dumpster on our campus at no charge. We didn't receive any payment from the sale of the paper, so we decided that we would also collect aluminum cans. In this way, our Science Club could earn cash for items we wanted to buy for our classroom.

Before the bins arrived, we decided on a plan to share our recycling idea with the rest of the school. Since we had the principal's approval, it was easy for me to ask teachers at a faculty meeting for permission to put a bin in each room—if students took responsibility for emptying them.

The Science Club decided that its members would go to each classroom and give a short speech about the recycling program. Then the class would vote whether or not they wanted to participate. With each bin came a page indicating the kinds of paper that were acceptable and unacceptable. On a Friday afternoon, members of the Science Club, ready with a practiced speech, went out in twos and threes to give their speech. Their speech was a simple act of persuasion. It contained the what, why, and how the school could recycle.

All classrooms agreed to give the project a try. At first, the Science Club members went and emptied the bins, but after a few weeks, students in each classroom wanted to take out the paper. In the first year of recycling, we emptied the dumpster about twelve times. In the second year, about fifteen times. Also during the second year, we added dumpsters for glass and cardboard. The third year, local laws had changed and the community was required to recycle many items. Our school had already been in compliance for three years.

Now recycling seems like a natural part of our school plan. Our need to persuade was subtle. Once each classroom of students bought in to our idea, it was just a matter of thanking people often for their support and reminding them of the kinds of paper that could be recycled.

You might wonder about our money raising efforts. The first year we collected about $24.00 worth of aluminum cans. They were messy, smelly, and, as you might imagine, took up a lot of space in my classroom. But they were fun to crunch. In fact, during an all-school field day, we made can crunching a game. Unfortunately, enthusiasm for the project waned when students realized that recycling cans made minimal money.

Approaches to Problem Solving

Problem solving is a part of everyone's daily life. Every time you ask students a question, they face a problem. Can they answer the question? Do they have enough time to collect their thoughts? In science, few important questions can be answered quickly and with a simple or single answer. Most problems raise questions that lead to hypotheses, which, in turn, lead to still more questions.

Problem solving requires abstract thinking. It plagues students in all subjects. Why? Many teachers assume that student training in problem solving in one area will transfer to all other subjects. Such transfer usually does not happen unless students work specifically on the transfer. Math problems usually are solved in a linear method. You state the rule or equation, write the steps, and arrive (sooner or later) at a solution. Most science problems are not solved easily in a linear manner; they require divergent thinking. For example, instead of using a linear sequence to solve a problem, try mentally setting the problem in the center of a polygon or cube and approaching it from different angles or points of view for possible solutions (something like the Cubing activity in Chapter 5).

Problems can be solved in different ways. Teachers share the responsibility to introduce students to a variety of methods for solving problems. Students should know how to recognize a problem. They must be able to define the problem and create a list of possible solutions. Students need to know that one "right" answer to a problem doesn't always exist. Students need to explore the facts, opinions, and values involved in the problem-solving situation and identify the variety of consequences that may arise from different solutions. They should be aware that problems may reoccur. Many real life problems reoccur, and students should have many strategies to solve them.

Objectives After sufficient practice, students should be able to

- identify one procedure for solving problems
- solve problems related to science content
- work collaboratively to solve problems

Uses Developing ways of solving science problems by asking questions and then trying to answer them. Developing hypotheses and suggesting ways of testing the hypotheses. Reinforcing learning to distinguish (compare) facts and opinions, causes and effects.

Directions Two sets of activity directions are included below. This first set of directions focuses on the thought processes involved in the processes of problem solving. The directions should be used to get students to think about how they solve problems. After this introductory lesson, problem-solving experiences should be integrated into your regular curriculum. This is a large group activity. You might want to read *Breaking Through: Creative Problem Solving Using Six Successful Strategies* by Tom Logsdon before you start this activity. (Logsdon 1993)

1. Write *problem solving* on the chalkboard.
2. Ask students for possible definitions. Write their responses on the board.
3. Decide on a working definition from student responses, and have students copy this into their daybooks.
4. Write one or two science-oriented math problems on the board. Have students solve these problems.

5. Ask students to explain what were they thinking when they solved the problems. What mental procedure did they follow? Write these steps on the board.

6. Discuss student responses. Correct any naïve conceptions.

7. Order the steps students used and ask them to copy the steps into their daybooks. Explain that many math problems are solved by a linear process. Explain that linear means "moving in a straight line to solve a problem." Refer to the math problems just completed.

8. Continue by explaining that most problems have several possible roads to a solution. You have to look at the problem from several different angles or points of view to get ideas on how to proceed. An important component of problem solving is feedback. You try a method and assess how it is working. You may do this several times until you have an acceptable solution for the problem.

9. Ask students to write other ideas they think might work for solving problems.

10. Discuss different ideas that students develop. Write these on the chalkboard.

11. Order the ideas into a format for problem solving. Have students copy these general methods into their daybooks.

12. Review the definition of problem solving. Review the linear method of problem solving.

13. Ask students to compare the linear method to the classroom-generated method of problem solving. Remind students that problem solving doesn't always follow a written pattern or procedure. Sometimes solutions seem to "come out of the blue," but most problem solving comes from asking appropriate questions.

The following second set of directions gives students practice with problem solving in small groups:

1. Review the definition of *problem solving*.

2. Explain that working together in small groups can be important in problem solving. Groups can often solve problems that would be too difficult or time consuming for an individual. Groups may have more information and knowledge than individuals acting on their own.

3. Explain that students will work in small groups to solve a problem. They will receive a group grade and are expected to help each other. Point out that these groups are different from peer response groups, where each student shares written work for revision help.

4. Assign students to their groups. Groups of three to five students work well.

5. Remind students that as a possible method they can use the procedures for problem solving they have written in their daybooks.

6. Introduce a problem to be solved.

7. Have groups brainstorm, list, or use another prewriting activity to produce possible solutions to the problem or to ask questions that will lead to methods of finding a solution to the problem.

8. Ask groups to list the proposed solutions or questions in the order they should be researched. Each group should submit a copy of their list.

9. Provide students with enough time to research their questions and to reach a solution to the problem. If this is a long-term project, you may wish to collect progress reports.

10. Have groups create a rough draft describing their solution. Rough drafts should be shared with another group for peer review. Then each group should revise their draft.

11. Have groups decide the manner of presentation of their solutions. Choices might include a poster, comic, story, essay, skit, speech, song, invention, or experiment.

12. Provide time for groups to work on their presentations.

13. Have groups present their solutions in a large group setting.

Follow-Up/Assessment Working together in small groups is the most important part of this activity. In the complex society in which students live, cooperation is absolutely necessary. Students are solving problems at several different levels all the time. When you stop and make them aware of the process, you help them learn that a variety of strategies may be used to solve problems. The strategies students use depend on the questions asked. Have students solve simple problems and write down the mental procedures they followed. This activity can also be scored using a checklist.

Examples There are as many varieties of problem-solving activities as there are problems. Some activities are more complex than others. Some problems— *How do you change a light bulb?*—have simple solutions. Humans have struggled with others for a long time: How do we conserve the world's biodiversity and still maintain the quality of life to which we have become accustomed? Students should have experience with both simple and complex questions.

Provide students with some long-term examples that can't be solved immediately. Also provide them with fun problems, such as the following:

- Design a microscope that can be taken on underwater field trips.
- Design an aquarium that allows water to circulate freely, but keeps organisms separate.
- Invent a new light switch.
- Invent a machine that runs on batteries.
- Modify an existing public domain software game so that it demonstrates a science concept studied in this class.
- Be a _____ for a day. (Have students role-play various animals or objects, such as a drop of water, a piece of trash, a drop of oil. This encourages students to use the imaginative side of their brains.)

Problems that might last the whole year could be ones that bring your students into competition with others. They could enter Invent America, the Duracell® competition, Science Olympiads, or local science fairs. These all involve problem solving and offer rewards beyond classroom recognition.

Here are some problems that have kept many scientists busy. Why not try them with your students?

- Tourists revisiting the Adirondack mountains of New York State have found several lakes where fish or other aquatic life no longer exists. Because of this, tourists have left and the tourism industry is dying. What caused this? Have students choose one of these four points of view and research possible solutions: economic, political, personal health, or environmental.
- Landscapers have imported from a foreign country a beautiful "cover" plant for private use. Unfortunately, this plant has no natural predators/competitors in the United States, and it "escapes" from the confined environments of local yards into natural habitat. What can be done? (Good examples of this problem are purple loose strife, water lilies, and other exotics.)

Thinking and Writing at Higher-Order Cognitive Levels

5

*T*he activities and strategies in this chapter are designed to provide your students with practice in thinking about and expressing their understanding of science concepts at higher-order cognitive levels. During activities, students will think about meanings and points of view, express ideas, and develop understanding about scientific concepts.

These activities challenge students to go beyond the lower-order thinking and science processes of knowledge and comparison. In the Shades of Meaning activities, students will work on using words along a continuum of changing meaning. Concept Mapping will help students analyze their own change in understanding. After exposure to the Cubing and Fact and Fiction activities, your students should have a better idea of expressing different points of view. Open-Ended Essays will give them practice writing about their scientific understanding using different styles of expression. Cubing and Open-Ended Essays are complex exercises; they will need to be repeated often before students will complete them with ease. Fact and Fiction is a two-dimensional version of cubing. You can put any two points of view together and use them as a writing/thinking activity or assessment. The activities in this chapter can be used for short- or long-term writing projects.

Activities in this chapter include:

- Shades of Meaning
- Concept Mapping
- Fact and Fiction
- Open-Ended Essays
- Cubing

Shades of Meaning

In this activity, word lists are examined, grouped, and arranged in some order based on the meaning of the words. The purpose of this activity is for students to learn more about similarities and differences among words. It also is a way for students to practice analysis and classification at a very simple level. Students use a variety of thinking patterns and can create their own hierarchy system as they arrange and group words. Listening to students explain or write how they arrange words might give you insight into their thinking patterns. Working with words that have similar meanings may help students understand language used in a scientific context. This activity is a way for students to study a variety of similar words and build their science vocabulary. Students also gain experience in grouping, and learn to accept the groupings of others as valid.

Objectives After sufficient practice, students should be able to

- differentiate meanings of words used in a scientific context
- order groups of words by specific criteria

Uses Classification skills. Use of metacognition. Increase vocabulary. Discriminate finer meanings of words.

Directions This can be a group or individual activity.

1. Display a short word list—about five words—with similar meanings, and ask students to define them on a sheet of paper. Use word lists mentioned in the examples section or words generated by the Sensory Perceptions activity in Chapter 2.
2. Ask students to share the definitions they wrote. Keep asking for different definitions until you have all that students came up with. Discuss the richness of word meanings. Tell students that in this activity they will group words. Explain that when words are sequenced and specific meanings are used, students will be able to describe scientific observations with greater clarity.
3. Handout 5-1: Weather Words contains an example set of words and instructions for a process students can use to group the words. The reflections in Part 2 are an important part of the activity. Have your students try it out. If you are not doing a weather unit at the time, make your own list of words. Or better still, have your students brainstorm a list of words that fit the topic, issue, or idea you are currently studying. For any word for which

students do not know the "science" definition, they can consult dictionaries, glossaries, or you for help.

4. Have teams or individuals group their words.
5. Then have teams display their word groups, with the organizing group title word and their justification statements, around the room.
6. Allow time for teams to walk around the room looking at other teams' groupings.
7. When everyone has had a chance to look at the work of the other teams, have a short discussion on the diversity of the groupings. Ask students "Which group is the correct one?" Hopefully someone will say "They all are!" Use this moment to comment that the groupings are all correct, and for this reason scientists are careful when defining words, phrases, concepts, and theories. Scientists want people to clearly understand the meanings of the words they use to explain their work. Scientists also debate and argue meanings of words at conferences and workshops.
8. Collect and score work.

Follow-Up/Assessment Score papers for completion and accuracy of science content. For variation of this activity, after students have grouped the words, ask them to choose one group of words, write each word on a separate slip of paper, and then organize the words in a row using criteria that they choose. Once done, ask another student to look at the words in the row and guess what criteria was used to organize them. This is a higher-order thinking activity. You can also use this variation for a quick check of understanding or a focusing device at the beginning or end of an activity.

Examples Description words for concepts may be grouped and ordered. Words must be similar yet different by some degree. One group of words for a habitat might include *variation, range, change, selection, wide, broad,* and *narrow*. Another group of words used to describe populations might be organized around their comparative character: *some, nearly, completely, all, none,* and *few*. A list of words used in describing rock appearances might include: *dark, clear, opaque, milky, bright, glassy,* or *pale*. Encourage students to use a thesaurus: it will provide lists of descriptive words, and students will gain familiarity with a useful reference source.

During a weather unit, students used Handout 5-1 to enrich their vocabulary when they were forecasting weather. Figure 5-1 shows how one group completed the assignment.

Figure 5-1
Sample Word
Grouping
(completed
by students)

Part 1		
Group 1	*Group 2*	*Group 3*
Temperature	Humidity	Conditions
hot, warm, cool	moist, wet, damp	overcast, cloudy, sunny
cold, polar, tropical	humid, dry, tropical	clear, foggy, windy
		calm, mild, stormy

Part 2

Group 1 is arranged by decreasing temperatures: tropical, hot, warm, cool, cold, polar

Group 2 is arranged by increasing humidity: dry, moist, damp, wet, humid, tropical

Group 3 was divided into two groups when we did this part:
- *The first division is arranged by the amount of cloud coverage: clear, sunny, cloudy, overcast, foggy*
- *The second division is arranged by wind conditions: calm, mild, windy, stormy*

Concept Mapping

Concept mapping is a way to organize information on paper in a graphic form. What makes concept mapping different from other graphic organizers is that the map is completed by the student. Connecting lines between words will often have written descriptors that help define the relationship between the two words. No prearranged structure is given. Starting without a structure allows students with different learning styles to work without restraint.

Objectives After sufficient practice, students should be able to

- show relationships between concepts by constructing concept maps
- use concept mapping to show understanding
- interpret concept maps

Uses Assess prior knowledge. Pre and post test. Check for understanding at conceptual level

Directions This can be a group or individual activity. The activity described below is for beginners. As students become more adept with concept mapping, eliminate the cutting and pasting steps and have students write directly on their papers. Remember that not everyone is adept at visualizing their personal conceptual understanding on paper. Some students will be very frustrated with this activity no matter how many times they do it. The Shades of Meaning activity in this chapter makes a good prewriting activity for this one.

1. Provide students with a large sheet of paper, small strips of paper, glue, and scissors.
2. Have students quickly brainstorm a list of words or phrases that relate to your current content. The first time you do this, you might use a reading where students glean out the important terms, phrases, or ideas. Review the words and phrases so that students have a working understanding of each. A list that contains fewer than twenty words is a good starting list.
3. Have students write ten to fifteen words or phrases from the list on small strips of paper.
4. Using a large sheet of paper as a background, have students group the word strips in a manner that makes sense to them.
5. When students are satisfied with their groupings, have them glue down the word strips onto the paper.
6. Direct them to draw connecting lines between words, using arrows as needed, until the concept map they have created makes sense to them.
7. Students can write words on the connecting lines between words to clarify the map.
8. Have students share their concept map with a partner. On the concept map, ask reviewers to write questions regarding any connections they think might need more explanation.
9. Collect concept maps with reviewers' comments for scoring.

Follow-Up/Assessment Scoring concept maps can take time. One teacher has a way of using student power to avoid hours of grading maps. She has students make a pre- and post-concept map during a module. Then using both maps, students write a short comparison paper on the following topic: What do I know now that I didn't know when I started? The teacher collects and scores these papers.

Another scoring method you might use is to give points for the appearance of an important items and another set of points for appropriate use of connecting words. You may also have students examine one another's maps.

Examples Figure 5-2 shows an example of a concept map from the weather unit.

Figure 5-2
Sample
Concept Map

Pre-lesson

Post-lesson

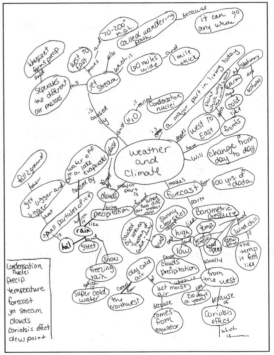

Fact and Fiction

Fact and Fiction is an activity where students write from two different perspectives or points of view. They also write in two very different writing styles. The fact part uses the lower-order thinking skill of description. Students show their knowledge and comprehension capabilities when writing in this style. The fiction part makes use of students' ability to synthesize information. This activity can be done as a summary for a unit, or at the end of a long-term investigation.

Objective After sufficient practice, students should be able to write about science concepts or an experience in two different writing styles—factual and fictional

Uses Writing from multiple points of view. Using description and synthesis. Using complete writing process. Summative assessment.

Directions This is an individual activity.

1. Discuss the difference between *fact* and *fiction* with students. Work with students to develop working definitions for each word. Have students write the definitions in their daybooks.
2. Explain that for this activity students will be writing from two different perspectives—one factual, and the other fictional. Have students list the characteristics of each kind of writing. Then have them list how they would show their understanding or learning using these two writing styles. These lists will become the basis for the scoring guide.
3. Design a scoring guide with students before they begin writing.
4. Have students write in class if the assignment is intended as an exam. Otherwise, you can assign the writing as homework.
5. Collect student writing for scoring.

Follow-Up/Assessment Score student writing according to the scoring guide. Remember, as you score, that this is first-draft writing. Student writings will be very different for each point of view, which is why these two particular styles were put together. You could also use any two different writing styles described in the Cubing activity in this chapter.

Examples You might use this idea for any number of essay exams. Give students a conceptual question. Then ask them a fictional one in which they have to apply the concepts.

From an interdisciplinary weather unit, the following two essay questions (out of four) were asked on a test. Students had time to prepare rough drafts. They wrote a final draft in class.

Fact What are the forces that drive our weather? Discuss how the sun, earth, air, and water cycle combine to create our daily weather.

Fiction Describe your life as a water molecule trapped in the water cycle. No rest for you in a water table or trapped inside the earth's crust. What are some of your adventures?

During a unit on energy transformations, you might ask the following questions.

Fact Describe the processes and major scientific concepts that are part of a hydroelectric power generating plant. You can use diagrams and pictures from your investigations.

Fiction Suppose you could step inside a turbine used in the generation of hydroelectric energy. Describe what could happen in one 24-hour period.

One Teacher's Story

Fact and Fiction

What did I learn by having my students write fact and fiction papers after a semester-long life science project? Rewards for me were threefold.

- *My objective of teaching careful observations was easily measured.*
- *My students' observations over the four and a half months of the experiment became more sophisticated.*
- *Students' predictions in the next week were more detailed.*

During the project, students observed and recorded the changes in a sealed jar to which soil and water had been added. The experiment was simple in materials and planning—a transparent jar, a little soil, a little sun, and a little time. Students enjoyed the experience. They wrote detailed descriptions on how they would conduct this experiment again.

Evaluating the project allowed me to assess student achievement from several different angles. I had written observations. I had written predictions. These became more sophisticated during the semester. I had short essays on changes. I had wonderful creative stories—not just the facts.

Figure 5-3 shows the final assessment questions students answered to show their competence in this experiment. In completing the Fact questions, did students incorporate first semester content and apply it to their discussion of their experiment? Yes, they did. And in completing the Fiction part of the assessment, could students use their observations to express another point of view? Yes, and I appreciated the imaginative ways they incorporated the facts into their stories.

Sample Final Assessment Questions for Preceding "One Teacher's Story"

Figure 5-3
Sample Fact or
Fiction Assessment

FACT

1. What happened in your mini-habitat? Summarize your observations in a sequential manner.
2. What did you learn as an outside observer about your mini-habitat? Describe any problems you encountered and how you solved them.
3. If you were going to do this experiment again, what would you do differently?

FICTION

Imagine that something in your jar could tell you its story. Write that story. Remember that your story must be based on your observations.

Open-Ended Essays

Mention essay questions and most students and teachers will give you a negative response. Essays have been the nemesis of science teachers for many reasons. One is that they are difficult to grade. Students have panicked at the thought of them because often they are asked to write essays only during tests. They are expected to interpret the question correctly, have a vast store of knowledge, write well, and do it all in less than an hour while answering many other questions as well. Many of the stresses involved with essay-question response writing are dispelled in this activity.

Open-ended essay questions do not have one particular correct answer. They are asked in a traditional format: first, set the focus of the question; second, give the directions for writing. Student responses may vary from a paragraph to many pages, depending on the complexity of the question. Open-ended essays are not necessarily included with several other questions in an test. They can stand alone and can be used for a variety of purposes.

For teachers, writing appropriate open-ended essay questions is a process that can be learned. The traditional essay question that says "Tell me in writing the five ideas about this concept" is boring for students to write and for teachers to grade. The same question also could be asked in a simple fill-in-the-blank or short-answer question. When you ask students to write in an expository manner, you should access their higher-order thinking skills rather than requiring simple memorization or recall of information. Open-ended essay questions can be fun, while they allow students another venue to show their conceptual understanding.

In open-ended essay questions, different response writing styles can be chosen by the teacher. Many writing styles suit the science classroom, including descriptions, comparisons, narratives, fiction, problem solving, analysis, synthesis, and persuasion. (Freedman 1994b) Collaboration with a language arts teacher will allow you to explore many other styles. When you create a scoring guide to go with the essay, the writing style should be part of the score. Another part of the score would be the appropriate use and depth of content. A third part might be style and usage.

Objectives After sufficient practice, students should be able to

- analyze the writing prompt for directions
- organize material for writing an essay
- write, using a particular writing style, about any science topic, idea, issue, or concept

Uses Formal assessments. Writing at higher-order thinking levels. Using multiple writing styles. Analysis.

Directions This activity should take place over several days. Many parts can be assigned as homework. It is an individual activity, with some small group work as well. The purpose of this writing activity is to give students experience in responding to essay questions and to give you a chance to develop a scoring guide for the questions. For this activity, students need time to prewrite, draft, revise, and rewrite their essays before they are graded. You will have to choose in advance what writing style you want students to use. You can use the prompts from the Cubing activity in this chapter as starters. Then you will have to give specific directions for writing.

1. Explain the process of writing an open-ended essay and how it is different from other essays your students might have written.
2. Distribute Handout 5-2: Practice Essay. Discuss the question in detail, making sure students understand the directions. The first few times students complete an open-ended essay activity, you might want to have the class brainstorm a list of the concepts related to the essay. Afterwards, students should be able to do the prewriting on their own.
3. Students prewrite. This can be homework.
4. Have students read their rough drafts to a partner. The listener should make comments on the writing style and the use of content. Students self-edit and seek help from you. As students are reading to each other, circulate around the room, making additional comments. This is one time when reading over the shoulder is highly acceptable.
5. Provide class time for students to write the final draft of the essay. They may use their rough-draft and prewriting materials at this time.
6. Collect the completed essays for scoring.

Follow-Up/Assessment Score and give credit for prewriting, rough draft, and final draft. A simple checklist with points could be used for this part: students can fill out the checklist themselves. Figure 5-4 contains some guiding questions that you can use to score the final draft.

General Assessment Area			
Conceptual Understanding	Content/Knowledge	Critical Thinking Process	Communication
What are the big ideas?	What are the facts, illustrations, descriptions, and examples students should have internalized while studying the big idea?	What kinds of thinking processes is the learner engaged in while answering this open-ended question?	How well can the learner communicate? Has she paid attention to grammar, style, and usage?

Figure 5-4
General Scoring Guide

Reprinted with permission from *Open-ended Questioning: A Handbook for Educators,* Addison-Wesley Publishing Company, 1994.

Using different writing formats for open-ended essays assesses different thinking skills. You can assess different thinking skills using the same concepts and content knowledge instruction just by changing the directions and modifying the question a bit. Figure 5-5 in the following example shows how questions can be altered to assess different thinking skills.

Examples The questions in Figure 5-5 were developed to assess different levels of thinking about one science concept. Each of the open-ended questions in the example is assessing conceptual understanding. The ways in which students would respond to each question and the scoring of these responses are different. The three scoring guides reflect the difference in the type of thinking that goes along with the responses to each question.

Figure 5-5
Sample Essay
Questions and
Accompanying
Scoring Guides

Concept The cell is the basic unit of life in plants and animals.

Writing Style Comparison

Prompt Imagine that you have the ability to miniaturize yourself so that you can travel freely around inside different plant and animal cells. Spend some time examining both plant and animal cells from the inside.

Directions Compare the two cells. What did you find that was similar? What differences did you discover? What surprised you?

Concept: The cell is the basic unit of life in plants and animals. Writing Style: Comparison	
Score	**Criteria**
5	Eight or more cell processes used in the comparison Ten or more cellular parts used in the comparison Student uses several different comparison strategies* Paper is well organized, coherent, and clear
3	Five or more cell processes used in the comparison Seven or more cellular parts used in the comparison Student uses at least two comparison strategies* Paper is organized, mostly coherent, and clear
1	Three or more cell processes used in the comparison Five or more cellular parts used in the comparison Student uses at least one comparison strategy* Paper has weak organization, sometimes incoherent, and not always clear

* Comparison strategies: side by side, lesser to greater, simple to complex, one-to-one, sequencing, and ordering by properties

Writing Style Problem-Solving

Prompt Imagine that you have the ability to miniaturize yourself so that you can freely travel around inside an animal cell. Spend some time examining different animal cells from the inside. A cancer has come in contact with the animal cells you studied. Which cells might be better suited to repel or destroy the cancer? Can you keep the cancer out of your cell or destroy it once it invades? How?

Directions From your knowledge and research, design a solution to help the cell keep out or destroy the cancer. What are the problems cells encounter in keeping foreign invaders out or in ridding themselves of invaders that get inside them? What are the probable ways a cell might defend itself against invaders? What are the advantages and disadvantages of the different methods of protection? Which solution do you think is best, and why? Draw a cartoon of what is happening at the cellular level that shows your solution.

Concept: The cell is the basic unit of life in plants and animals.	
Writing Style: Problem Solving	

Score	Criteria
5	Solution presented in cartoon form A solution is presented Evidence supports solution Paper is well organized, coherent, and clear
3	Solution is presented in cartoon form A solution is presented Evidence doesn't always support solution Paper is organized, mostly coherent, and clear
1	No cartoon Solution may be vague If evidence is presented, it doesn't necessarily support solution Paper has weak organization, sometimes incoherent, and not always clear

Writing Style Evaluation

Prompt Imagine that you have the ability to miniaturize yourself so that you can freely travel around inside different plant and animal cells. Spend some time examining both plant and animal cells from the inside out.

Directions After your journey inside both plant and animal cells, choose the one you would rather be. State your judgment and support it with reasons and evidence. Include as least three reasons that are supported by research in your response.

Concept: The cell is the basic unit of life in plants and animals.
Writing Style: Evaluation

Score	Criteria
5	Three or more cell processes used in evidence or reasons Three or more cellular parts described in evidence Judgment statement clear and concise Three or more reasons stated logically Evidence that supports each reason
3	At least two cell processes used in evidence or reasons At least two cellular parts described in evidence Judgment statement mostly clear Two or more reasons stated logically Evidence that supports each reason
1	One cell process used in evidence or reasons Cellular parts mentioned in evidence Judgment statement not clear One or more reasons stated but may not be logical Evidence supporting each reason, but may not be logical

Cubing

Cubing is a discovery method that involves the detailed analysis of a topic, issue, idea, or concept. Students choose a topic and write about it from six different perspectives: describing, comparing, associating, analyzing, applying, and arguing for or against. Note that these six perspectives are writing styles, not philosophical points of view. Students may form new and original connections about a topic from exploring a topic from various points of view. This activity employs higher-order thinking skills. It is one of the most complex writing activities included in this book.

Objectives After sufficient practice, students should be able to

- brainstorm six aspects of a topic
- write about a topic from six different perspectives

Uses Writing from multiple perspectives. Synthesis of ideas. Divergent thinking. Special projects. Summative assessment. Organization of materials.

Directions Before doing this activity, students should have experience in writing from several different perspectives. You might have students practice writing from different perspectives until you feel they are ready for cubing. One way to practice is to construct a 20-cm cube from tag board. Write one of the six perspectives on each side of the cube. The cube can be rolled and the perspective that appears on the top is the one that students write from that day. When students are ready, use the directions below as an individual activity.

1. Write the topic, idea, issue, or concept on the board.
2. Distribute Handout 5-3: Focus on Cubing. Explain that students will be writing about the topic from six different perspectives. They should be familiar with the cube by now, but you may want to discuss the different perspectives with them.
3. Have students write their ideas that relate to the topic alongside each square. If they need more room, they can write the cubing focus questions for each perspective in their daybooks and their ideas immediately after each focus question. The focusing questions may also be written on the board. If this is the first time your students are doing this complete exercise, you might want to structure the activity so that they are moving from the lower-order to higher-order styles. Their ideas will flow easier by the time they get to the analysis and argument styles.

4. After students have finished prewriting, ask them to write a general statement about the topic. This statement will become the main point or topic sentence of their essay. Students can write one statement for each perspective and then synthesize these statements into one idea.

5. Discuss students' prewriting. Correct any naïve conceptions.

6. Ask students to write a rough draft. Remind students that their paper should include all six perspectives. If you are using this as an exam, you might give students the topic ahead of time and have them come to the exam with their prewriting already completed.

7. Divide the class into response groups of three to five students.

8. In response groups, have each student read aloud his or her rough draft twice. During the first reading, other members of the groups only listen. During the second reading, group members fill out response forms. Handout 5-4: Cubing Response Form is a sample response form you can start with. Response forms should be thoughtfully filled out by the student reviewer and thoughtfully read by the author. Remind students that when someone has listened to their paper twice and cannot pick out the main idea, they may need to seriously look at their writing. You will want to provide advice freely after critiques are handed back. Remember that an atmosphere of trust is vital in this part of the writing process.

9. Direct students to use response forms to revise their rough drafts. Remind them that they should carefully read the response forms filled out by their response group members, but that they have final say in the changes they make in their papers. They also should self-edit their papers for style and usage. This could be homework.

10. Have students prepare a final draft of their work. As they put together the pieces to turn in, have them fill out Handout 5-5: Cubing Checklist and include it as a cover sheet. As students gain experience with cubing, you will want to get student input when constructing the checklists, but the first time students do this activity, provide one for them. Students should organize their work according to the checklist.

11. Collect prewriting, rough draft, critique sheets, final draft, and a completed checklist from each student.

Follow-Up/Assessment This activity should produce polished essays that are rich in detail and are focused on the topic, issue, idea, or concept. The essay should be evidence of student understanding and competency in sharing ideas in a written format. Using a checklist will make scoring easier for you.

You can use the following activity as a prewriting activity, as a review, or as a preview to a new unit. Explain that you are going to give each group a perspective from the cube and the group will have two to four minutes (depending on your class) to write down everything they can think of about a topic using the perspective they are given. Then they will have one to three minutes to organize their thoughts. Afterwards, one member of the group will speak to the rest of the class about the topic in the assigned style. You might even designate the groups and their assigned writing styles ahead of time. Talking about what students might write about is a worthwhile prewriting activity. It might just be a little bit fun as well. No grades or competition here, just participation points.

Another way you might work with this activity is to have all students write from one perspective. Switch perspectives until everyone writes appropriately from all six. When you feel students are ready to go on, choose a topic/issue/concept as the focus. Divide the class into six groups. Each group is then responsible for writing from their prearranged perspective. Have groups give an oral reading of their perspective. Discuss.

Examples In one middle-school science course, students had very little experience in writing from several different perspectives. First, they had to learn to observe from more that one point of view. Then they could write about it. Over the course of a school year, they gained experience from this activity.

Activities for a Change of Pace

6

*T*he activities and strategies in this chapter are designed as options to traditional learning and assessment methods. They focus on learning styles other than analytical/mathematical and verbal/linguistic. Because these activities focus on learning styles that may not be typical of a science classroom, red ink and high expectations the first time they are used may act as a dam to ideas—and idea expression that is a crucial part of learning and assessment.

Some of these activities may be new to your students in regards to science content. Be prepared to try these more than once. The first time you try them may not give you the kinds of results that practice will provide. Use the activities in this chapter as both learning and assessment activities. Score the ones that are useful for deciding grades, and use the others as learning tools. Several activities require students to use the complete writing process—prewriting, composing, editing, and sharing/publishing. You can choose to do one or more parts of the writing process for your assessment. Many of these activities could serve as long-term projects.

Activities in this chapter include:

- Literary Devices
- Science Acronyms
- Word Pictures
- News Briefs
- Letters
- Posters
- Who or What Am I?
- Comics
- Skits
- Stories

Literary Devices

Literary devices, such as similes, metaphors, and personification, encourage students to think creatively about content. Both metaphors and similes compare two things. *Similes* compare things explicitly, using the words *like* or *as.* In the sentence *We were fascinated by the euglena's wormlike movement,* the simile *wormlike* tells us that the euglena's movement resembles that of a worm. The simile enables us to compare something possibly less familiar (the movement of a euglena) to something more familiar (the movement of a worm). *Metaphors,* on the other hand, imply comparison without using words such as *like* or *as. The field of flowers was a rainbow of color* is an example of a metaphor.

Personification treats ideas and objects as human, giving them human characteristics. An example would be *The daisy is smiling at you.* Usually personification is not an acceptable practice in science. But when it comes to literary devices, it can help students describe unfamiliar science terms and concepts in familiar words or phrases. While you should not encourage the use of personification, students should be able to recognize when it is being used in reading materials.

Literary devices use divergent-thinking processes. They change the point of view of student writing. If students can imagine what it would be like to be a test tube or describe a day in the life of an organism in a micro-habitat, they might understand science concepts more easily. When they use these devices, students are creating mental pictures. Imagery is an important component of thinking processes.

Objectives After sufficient practice, students should be able to

- write creatively about science topics using metaphors and similes
- recognize personification.

Uses Interdisciplinary connections. Descriptions. Writing from a different point of view.

Directions Each literary device can be introduced individually, or they may all be introduced at the same time. The directions are written for use when introducing the choices at the same time, but the same directions would apply when introducing them individually. This is an individual activity.

1. Write *personification, metaphor,* and *simile* on the chalkboard.
2. Share several examples of each literary device.
3. Ask students to define each word. Write these definitions on the board.
4. Work with students to develop a working definition for each word. Have students copy the definitions and examples into their daybooks.
5. Write a topic on the board.
6. Ask students to cluster or list possible ideas for the topic.
7. Have students use focused free writing to several paragraphs on the topic, using the literary devices you have introduced.
8. In response groups, have students share their writing for comments.
9. Have students revise and rewrite their papers.
10. Collect student work for scoring and publication.

Follow-Up/Assessment Using a 100-point scale, you might assign points to the Literary Device papers as follows: prewriting—20 points; focused free writing—20 points; revision—30 points; use of content—30 points.

Examples Use words and phrases such as *pretend, imagine, describe, I wonder,* and *what if* when giving students practice with literary devices. The following are some examples to get you started.

Chemistry
- Pretend you are a piece of laboratory equipment stored in a drawer. Write a paragraph about what you think it might be like. Choose from the following list of equipment: dirty test tube, clean test tube, Bunsen burner, safety glasses, pipette, ring stand, graduated cylinder, beaker, Erlenmeyer flask, or clamps.
- What if you were a copper atom? Describe your day in a copper sulfate solution.
- Imagine that you are a nickel-cadmium battery. Describe your life.

General science or biology
- Suppose you were _____ in our classroom. (Choose from the following list to fill in the blank: the aquarium, a fire extinguisher, a fire blanket, the plants in corner, the bulletin board, a student's desk, or your desk.) Write a paragraph telling about a day in your life.
- Choose a protist from those being studied. Follow it around for a day. Describe what a day in a drop of water must be like for the protist you chose.

Earth science

- What if you were a cloud, a hailstone, an earthquake, a grain of sand on the beach or in a river. Describe what your day would be like.
- Suppose you were a watershed. Describe what it would be like.
- I wonder what it would be like to be a molecule of water. Describe your life in the water cycle.

Physics

- Imagine you were a beam of white light. Describe what it would be like to travel through a lens, a lake, or a prism.
- Pretend you are a molecule of water trapped in a balloon. Describe what happens to you as the balloon is heated and cooled.

Science Acronyms

Science acronyms are a variation of traditional acronyms. The name of an organism, topic, person—or just about any noun—is placed vertically down the side of a page, one letter per line. Each letter begins the first word in a statement or phrase that gives information about the word or topic.

Students might find this sort of activity a welcome change from brainstorming or clustering a topic. In order for an acronym to make sense, students must understand the principles, concepts, and processes connected with the word. This type of assessment combines the thinking processes of knowledge, description, and synthesis with language use.

Objectives After sufficient practice, students should be able to

- write acronyms accurately using current science subject matter
- use acronyms as a means of expressing scientific knowledge

Uses Synthesis of ideas. Assessments. Research focus. Defending a point of view.

Directions This is a small group or individual activity.

1. Introduce a topic. With the class, create a list of terms related to the topic that could be used in the acronym activity.
2. Have groups or individuals choose a word from the list.
3. Tell students to write their word vertically on a sheet of paper. Each letter should begin one horizontal line.

4. Explain that each letter should begin a statement or phrase that contains information about the topic.

5. Provide time for students to work on their acronyms in class, or assign them as homework.

6. Acronyms should be shared in small groups of students who chose the same word. As each student reads aloud his acronym, the group may challenge any line if they think it is incorrect or inappropriate. If a student can successfully defend her writing, the line remains. If the statement can't be defended successfully, it must be changed. Because of the small group setting, this type of assessment gives students experience with a nonthreatening method of defending their work.

7. Have students revise their acronyms and prepare them for visual display.

Follow-Up/Assessment Acronyms make good homework assignments and examinations. A scoring guide should be posted for students to see. See the sample in Figure 6-1.

Score	Criteria
5	Each line contains appropriate material Several conceptual ideas included All challenged lines defended accurately
3	All lines completed Mixture of facts and concepts represented, not all appropriate Some challenged lines not defended accurately
1	Some lines unfinished Only facts Challenged lines undefended

Figure 6-1
Sample Acronym Scoring Guide

Examples You may wish to start this activity early in the year—perhaps the first day—and have students make an acronym using their name. They could use words that describe what kind of scientist they are or would like to be. Or they could use words that describe themselves, their hobbies, or their talents. Do not score this assignment, but use it as a way to get to know your students.

One group of students created the acronym in Figure 6-2 while studying plate tectonics. The words were displayed on a bulletin board with paper strips that could be removed. When someone had an idea for a line, they discussed it with the teacher, and, with approval, wrote the sentence or phrase on the strip of paper and posted it.

Figure 6-2
Acronym
(created by a
group of fresh-
man earth
science students)

Sample Student Science Acronym—Earth Science

Pangaea was a prehistoric land mass that eventually formed present conditions.

Lifts land in some places.

Always moving land somewhere

Tectonic is from the Greek, meaning "builder."

Earthquakes happen at plate boundaries.

The movement of the plates is predictable.

Earth is sometimes buried by plate movement.

Crustal movements cause earthquakes.

Trenches come from subduction.

Oceans contain a ridge where plates separate.

New land is formed by volcanoes.

Iceland is along a plate boundary.

Continental drift theory explains the movement of continents.

Spreading occurs at mid-ocean ridges.

Word Pictures

The word pictures in this activity use letters and/or pictures to combine art and science in a creative way to illustrate a science concept. They can be made when students know the concept behind a word and can express the concept in a picture within a word setting. Students need a high level of concept understanding to create words as pictures. Using word picture assignments with ESL and LEP students will help them increase their science vocabulary.

Objectives After sufficient practice, students should be able to

- identify science ideas and concepts using word pictures
- design their own word pictures

Uses Illustrate science concepts. Small group work. Integrate art with science.

Directions This activity may be done individually or in small group.

1. Show students several examples of word pictures, such as those in Figure 6-3.
2. Discuss how, in each example, the word picture demonstrates the concept.
3. Ask students to create their own word pictures.
4. Collect and display.

Follow-Up/Assessment Score for completion only.

Examples Figure 6-3 provides several examples of word pictures. Most words that are topics are possibilities for word pictures. A seventh grade class studying integrated science used the following words to make word pictures: *control, variable, biome, air, water, soil, habitat, adhesion, cohesion, solvent, precipitation, condensation, percolation, evaporation, opaque, transparent,* and *translucent.*

Figure 6-3 Word Pictures

News Briefs

News briefs are short—less than five paragraphs— science stories that are meant to give the essential information about a newsworthy item. Students should be exposed to news briefs from a variety of sources. The format and concise writing style of Science News Letter is well suited for reading aloud. NSTA's journals include science briefs. They should be read as examples of how writers condense information in an interesting manner.

Ask students to write news briefs about what's happening, different points of view, the latest science ideas, or even imaginary events that students would like to see happen. News briefs relate science to the everyday activities and lives of students.

Objectives After sufficient practice, students should be able to

- write a short summary of a current topic for an audience of their peers
- interpret events in science from reading news briefs

Uses Informal assessments. Complete writing process practice. Short essays. Research projects.

Directions This activity uses the complete writing process. You may wish to review Chapter 1 before having students do this activity.

1. Write a topic on the board. It may be one that you have chosen, or one that students are personally interested in.
2. With the class, cluster topics for ideas.
3. Have students research the topic.
4. Ask students to choose their audience. They may choose peers, parents, younger students, a local newspaper, or any other audience.
5. Have students choose a format for their article. They might choose cause and effect, comparison, fact and opinion, persuasion, problem solving, information, or any other format.
6. Have students write a rough draft of their news briefs using the information they gathered.
7. In small groups, have students review each other's rough drafts.
8. Ask students to revise and edit their articles. A computer lab would be helpful at this stage. You might schedule student conferences during this time to review work.
9. Students should repeat steps 6 and 7 many times until their work is ready for publishing. Then have students write their final drafts.

10. Students should prepare any accompanying artwork or photographs.
11. Have students submit their finished article for publication.

Follow-Up/Assessment Using a 100-point scale, you can award the following points for each section of this assignment: cluster—15 points; research—15 points; rough draft—20 points; revisions—35 points; final draft—15 points. This writing activity could culminate in a science newsletter, cross-age reading, a focused-issue or topic bulletin board, or submissions to a local newspaper.

Examples This activity could be used almost anywhere you would like students to write a short, finished paper.

A theme newsletter using the format of Science News was the center of a unit on the cell for one class. Instead of a final unit exam, students created a newsletter for which they researched, photographed, created artwork, revised, rewrote, and published. Other newsletters might focus on laboratory safety, scientists of the twenty-first century, space shuttles, or many other topics.

A strategically located question box could be a source of topics. Students would answer the questions in news briefs. Briefs could be placed on a computer bulletin board accessible to other schools, or in a school newsletter, a local paper, or even a journal.

Letters

Having students write letters is one way to shift the audience from you, the evaluator, to someone else. When students write to teachers, they assume that teachers are experts, and thus leave out basic descriptions that are needed for understanding. Students will write what they think their teacher wants to read. If students are asked to write to the principal, their parents, or to almost anyone else, they often write more creatively and with more details. If students can accurately explain a complex topic to someone else in a fluent fashion, they show their understanding of the subject.

Just because this letter is mimeographed, doesn't mean I'm not sincere.

Mae West

Objective After sufficient practice, students should be able to explain scientific information to someone who is not familiar with it.

Uses Informal assessments. Change audience for an assignment. Practice with letter writing. Synthesis of ideas.

Directions These are general directions for using this individual activity.

1. Explain to students that they will be writing a letter to someone about one of the science topics they have studied. You can choose the intended audience and the possible topics, or allow students to make these choices. The completed letters may be sent to their intended recipients.
2. Write the possible science topics on the board.
3. If you have sample letters written by students, read these to the class. Discuss.
4. Design a scoring guide or checklist.
5. Have students copy the scoring guide into their daybooks. Or you might copy the guide and distribute it.
6. From the list of science topics on the board, have students choose the one they will address.
7. Students should use one of the prewriting activities from Chapter 2 to help them gather ideas.
8. Have students organize their ideas and write a first draft of their letters.
9. Depending on the scoring guide, students will then submit their rough drafts for scoring or editing, rewrite it, and then turn in their final letters.

Examples In a high school science class studying the history of science, students wrote unmailed letters to inventors of various scientific apparatus. They told the inventors how their inventions are being used today and described other inventions that have been developed because of the initial inventions. Chemistry students wrote unmailed letters to the discoverer of various elements. Students told the discoverers how the elements have contributed to modern times.

A seventh-grade class, working on a unit about soil and its inhabitants, wrote undelivered pre- and post-unit letters to the principal about earthworms. Some of the pre-unit letters included drawings of earthworms and information on how an earthworm benefited the soil. Post-letters included pictures of soil that had been worked over by earthworms compared to soil that had no earthworms. Since internal anatomy activities were a part of this module, some letters also included detailed drawings of the circulatory system of earthworms. Pre- and post-unit letters were compared for the change in knowledge.

Posters

Posters give students experience in combining many of their learning styles. They must use divergent thinking and imagery to create posters. Posters provide a visual representation of student knowledge. Posters can tell the facts or use persuasion. They can enhance a room's decor and make great bulletin boards. Posters also make wonderful interdisciplinary projects.

Objective After sufficient practice, students should be able to present their knowledge using visuals

Uses Informal assessment. Integrate art with science. Synthesis of ideas.

Directions These are general directions for creating posters. This can be a small group or individual activity.

1. Present students with the topic for their posters. Brainstorm the criteria on which they will be scored.
2. From students' brainstorming, create a checklist of critical items to include.
3. Have students brainstorm possible ideas to include on their posters.
4. Have students submit a preliminary idea sheet for review. It should be compared to the critical items list and awarded points. This will enable you to correct any naïve conceptions before students begin their final work.
5. Allow sufficient time for students to complete their posters either at home or in class.
6. Collect final posters for evaluation and publication. You may wish to provide several bulletin boards for displaying posters. Provide students with class time to describe their poster to the class.

Examples One class created pro-and-con posters for a drug unit. Other students created advertisements for an imaginary sound device. Another group chose a special-interest science group they would like to join and created a poster to recruit others to support the group's causes. Technology week, Earth Day and several other theme weeks in education are accompanied by posters that can be used as models.

Who or What Am I?

This activity is a variation of the childhood game Twenty Questions. Students write several statements to describe a person or an object. When these statements are arranged from most general to most specific, they form the basis of a Who-or-What-Am-I? description. Ordering, grouping, and categorizing are processes exercised while writing a Who-or-What-Am-I? description. When students are gathering information, they will look for the basic characteristics of the person or thing they are describing and then arrange their information in a particular pattern. This activity could be a precursor to working with dichotomous keys or exploring relationships in the periodic table.

Objectives After sufficient practice, students should be able to

- describe something in a number of clear, concise statements
- prioritize statements from the most general to the most specific

Uses Synthesis of ideas. Summative assessment. Research project. Classification and ordering process.

Directions This is an individual activity during the writing and a partner activity during scoring. The activity would be appropriate after a series of investigations or activities where students have explored a variety of characteristics that belong to different objects.

1. Provide time for students to choose, cluster ideas, and research a topic. Students should keep track of references.
2. Ask students to write several statements about their topic.
3. Have students arrange the statements in order from the most general to the most specific. Then have students number their statements in that order.
4. Have pairs of students read to each other, edit, and rearrange the statements, if necessary. You should act as a consultant to give pointers on the arrangement of statements. Have students write a final draft of their ordered statements for homework.
5. The next day, place student descriptions around the room.
6. Provide each student with two copies of Handout 6-1: Who or What Am I? Scoring Guide.
7. Have students read at least two Who-or-What-Am-I? descriptions and score them. Scoring sheets should be left with the descriptions.
8. Allow time for students to pick up their Who-or-What-Am-I? descriptions and read the scoring sheets.
9. Discuss those points students choose to share.

10. Collect completed checklists, clusters, research notes, statement pages, and scoring sheets for scoring.

Follow-Up/Assessment Use a scoring guide similar to the one in Figure 6-4 to score each section of this activity. In fact, you can display it ahead of time for students' own evaluations. Remember that you can change any item in the scoring guide to suit your students' developmental level. For younger students, you might read each statement of a Who-or-What-Am-I? description aloud and have them draw what they think it is. Animals and elements make excellent Who-or-What-Am-I? subjects. Topics, issues, and concepts might also become a Who-or-What-Am-I? topic. After reading several different Who-or-What-Am-I? descriptions, ask students to identify patterns in the way the questions are arranged. The patterns they discover might lead to a general statement about hierarchical classification systems!

Score	Criteria
5	Ten or more appropriate defensible statements
	Statements are arranged from most general to most specific
	Statements include six or more different characteristics
3	Eight or more appropriate defensible statements
	Statements arranged from most general to most specific, but some statements not in the best order
	Statements include three or more different characteristics
1	Fewer than eight appropriate defensible statements
	Statements are mostly random in arrangement
	Statements include fewer than three different characteristics

Figure 6-4
Sample Scoring Guide for a Who-or-What-Am-I? Description

Examples Students in a biology class were asked to choose an animal and create a Who-or-What-Am-I? list of statements for an examination. They were given the information in Figure 6-5 to guide their research.

Figure 6-5
Sample Instructions for a Who-or-What-Am-I? Description

1. Choose one animal species from the five classes of vertebrates we have studied.
2. Research more information.
3. Write a complete description of your animal. You will be creating a What-Am-I? list of descriptive statements.
4. Use these questions as a guide to your research.
   - Where does it live (habitat)?
   - Are its senses well developed? How do you know?
   - Does it have limbs? How does it move?
   - Is it cold-blooded or warm-blooded?
   - Is it a herbivore, omnivore, or carnivore?
   - How does the animal behave?
   - What characteristics make this animal unique?
5. Organize your statements from the most general to the most specific and number each statement.
6. Bring your rough draft to school for revision.

Students chose a variety of species. One sample of the work students submitted is shown in Figure 6-6.

Figure 6-6
Sample Who-or-What-Am-I? Description

Sample Student Answer to Instructions in Figure 6-5

What Am I?

1. I live in the wild.
2. I am hunted for food by man.
3. All of my body systems are well developed.
4. I use all five of my senses.
5. I have two pairs of limbs.
6. I am warm-blooded.
7. My body has hair.
8. I am a placental mammal.
9. I am a herbivore.
10. I gallop or trot with powerful legs.
11. I have hoofed feet.
12. I have antlers.

What Am I?

Comics

The use of comics or storyboards evokes creative ideas in students. Comics can present learning in a less threatening manner than other methods of instruction. Students will remember cartoons or comics long after they have forgotten facts, topics, or memorized trivia. Creating comics uses three thinking processes: idea synthesis, procedure writing, and communication.

Objectives After sufficient practice, students should be able to

- use comics or story boards to demonstrate conceptual understanding
- create comics to explain science concepts

Uses Integrate art with science. Synthesis of ideas. Check for understanding.

Directions These general ideas for using comics are not arranged in a specific order.

- Place several science-related comic strips around the room or on a bulletin board.
- Ask students to bring to class science-related comics from the newspaper.
- Discuss how science can be learned through comics.
- Ask students to create comics of their own.

Follow-Up/Assessment Don't score students' art abilities or humor. Award points for completion. Score the science content and the demonstration of conceptual understanding. You might plan a cooperative assignment with an art teacher, who could critique the artistic techniques.

Examples Laboratory safety is so important that teachers constantly refer to proper safety techniques. Why not have students create comic posters that depict safety? In this way, safety procedures can be reinforced in a pleasant manner. One teacher had a student who created a "safety turtle" that was used all year as an example of good laboratory procedures.

NOTE: Comics can be used to show comparisons. Posters on drug use, environmental issues, or other topics could be used to assess students' knowledge. Simple directions also might be done as comics.

Skits

Skits or short plays are another way for students to display their understanding of content. Most students enjoy being part of the production of a skit or short play. Their talents may be in writing scripts, editing, directing, or creating props. A skit can be as short as a few minutes, or as long as the imaginations of your writers will take it. Class productions will vary according to the sophistication of your writers.

Some of the benefits of producing skits might include building group cohesion, creating an atmosphere of mutual respect and trust, engaging students directly in content, providing inspiration for viewers to peruse scientific research, and teaching others science concepts.

Plays and skits help students visualize processes and explore models that are not easily demonstrated in a classroom. They tap into several multiple intelligences—linguistic, spatial, musical, logical/mathematical, body/kinesthetic, and interpersonal and intrapersonal social intelligence. Skits are memorable and can help relate science concepts to students' lives. They make excellent cross-age activities.

Objectives After sufficient practice, students should be able to

- demonstrate their ability to work cooperatively
- design, develop, and produce a skit illustrating a science topic

Uses Summative assessment. Cross-age activities. Involve large groups of students using multiple learning styles. Visualization of science concepts.

Directions The complexity of student skits depends on the time available, the intended audience, and the sophistication of your students. Simple demonstration skits can require only a small part of a class. A longer production could be a semester project. Most of the work can be done outside the classroom. Directions are for student creation and production of their own skits. This can be a large or small group activity.

1. Present students with a science topic and discuss the intended audience.
2. Have students divide into small groups for script writing. Explain that students may research the topic and write a play from their research. Or they may adapt one of their other papers, stories, essays, or research papers into a play or skit.

3. Ask students to brainstorm or cluster ideas for their skit.
4. Post a list of available resources. This list might include a timeline for performance, available equipment, consultants, music, recording devices, or supplies.
5. Have each group write a rough draft.
6. Have groups read their rough drafts to a response group for review.
7. Ask groups to revise their rough drafts based on their peer reviews. You might have teacher consultations while rough drafts are being revised.
8. Have students collect and make props and costumes, create a program of events, and send invitations, if appropriate.
9. Provide practice time for students to fine-tune their skits.
10. Have groups perform their skits for the intended audience.
11. Score student performances. You can videotape the performances and save these for viewing on the last day of school or the first day of the next year.

Follow-Up/Assessment You might want to involve theater arts teachers or language arts teachers with this assignment. If you plan a traveling production, make sure you have plenty of planning time. Permissions and other arrangements take time, but they are worth the effort. Finding a few other teachers to share the burden of responsibility will help.

Skits, puppet shows, or multiple productions might be planned around a special event. Design a scoring guide ahead of time so students know that participation, not innate acting abilities, will be scored.

Examples Simple classroom skits could take the form of puppet shows, mime, straight acting, or any other dramatic technique that students know. The best science skits are ones that demonstrate a process or activity that is hard for students to visualize. These might include cellular activities, weather cycles, or chemical reactions.

Possible topics include laboratory safety, the uses and life of a microscope, the cell, respiration, photosynthesis, famous moments in history, the life of an organism, the inner functions of a simple machine, the rock cycle, mutations of chromosomes, cellular division, chemical reactions, life as a cloud, electron bonding, life of an ion, DNA or protein synthesis, "Newton revealed," life as a gas atom, and life as a member of an endangered species.

One biology teacher uses two short plays to exemplify the light and dark reactions of photosynthesis. She said students always laugh when the ADP-H is carried into her closet for the dark reactions.

Many of the books in the *Nature Scope*® series include plays. They often are written for younger audiences, but middle-school and high-school students might also enjoy producing and acting out the productions.

Stories

Stories can become a favorite extra-credit assignment for you and your students. Students may write fiction or nonfiction stories for themselves, for younger or older students, or for a fictional audience. In order to write science stories, students must understand science content. They must be able to synthesize ideas into interesting contexts. When you ask students to write stories, you are giving them experience in combining synthesis and creativity. You are encouraging them to internalize and take ownership of content. To help students focus on story writing, read them several science-based literature books written for younger audiences.

Objectives After sufficient practice, students should be able to

- show creative use of science concepts
- write short stories

Uses Synthesis of concepts. Practice complete writing process. Cross-age tutoring.

Directions This is an individual activity.

1. Discuss with students the audience and science concepts to be used in their stories.
2. List the science concepts to be included in the story.
3. Have students cluster or brainstorm story ideas.
4. If there are any significant holes in the cluster, have students do research.
5. Have students write a rough draft of their stories.
6. In response groups, have students review each other's rough drafts.
7. Ask students to revise and edit their stories. You might have student conferences during this time to review students' work.
8. Have students write a final draft.
9. Have students prepare any accompanying artwork or photographs.
10. Collect each student's cluster, rough draft, response reviews, and final draft for scoring.
11. Have students submit their finished stories for publication. Ask students to share their stories with the intended audience.

Follow-Up/Assessment You can award points on a 100-point scale as follows: clustering—20 points; rough draft—20 points; revisions—40 points; final draft—20 points. Final-draft points might also include oral presentation credit. Provide students with a scoring guide ahead of time to focus their attention on components of their story that may need improvement.

You might arrange a cooperative assignment with a language arts teacher. Students are frequently asked to write short stories in those classes. Perhaps students can earn credit in two classes with this activity.

Writing stories for a younger audience can open the doors to cross-age tutoring. You might arrange a visit to a younger class where your students could present their stories and science demonstrations.

Examples One general science teacher had students write a story titled "Backstage with Your Favorite Rock Star." The story explained how sound, body systems, and rock concerts are connected. In an earth science class, students wrote about solar system journeys. Students in a biology class wrote cell tour stories for sixth graders. During a unit on light, students wrote stories on mirrors and lenses. These stories included trips to carnival fun houses, a mall, and their own rooms.

Poetry and Song

7

*T*he use of poetry encourages student imagination and increases language competency. Poetry extends science concepts into the realm of imagination. Writing poems can become an extension of classroom observations or can be used as a memorization device for important facts. Writing poems involves the synthesis of facts, concepts, and observations. Many of the prewriting activities in Chapter 2 can be used to collect information for poetic writing. Guided imagery is highly suitable for use with poetry because of its use of sensory information.

Song writing can be as simple as rhymed poetry set to music. For your auditory learners, science concepts expressed in songs will help them remember the facts.

Poems and songs are also easily published. Giving your students a public forum to show off their knowledge and earn positive recognition at the same time is an esteem-builder that should not be neglected. Poems and songs also make excellent cross-age sharing activities.

Activities in this chapter include

- Poems
    Bio Poem
    Cinquain
    Haiku
    Sonnet
    Limerick
- Songs

Poems

When students write poems, they use the higher-order thinking skill of synthesis. Students must have a conceptual understanding and quite a bit of factual knowledge to create different kinds of poems. Poems become conceptual word pictures of different scientific concepts. Using the various poetic forms included in this book will give your students different methods of consolidating their knowledge into discrete units. In several activities in this book, students are asked to write a general statement summary about the preceding work. A poem can be an alternative format for that statement. Writing haiku can lead to observing nature in a different way. Writing bio poems will give students practice with vocabulary. Writing cinquain and diamonte poems will help students pay attention to the structure of words. Limericks and sonnets expand the use of language.

Objectives After sufficient practice, students should be able to

- use various prewriting skills to develop ideas for poems
- use basic understanding of science concepts to create first-draft poems
- explore different styles of poetry
- share poems with peers
- rewrite poems for publication
- synthesize information into concise, comprehensible statements

Uses Writing for publication in school and community newspapers. Building observation skills. Increasing facility with science vocabulary. Broadening understanding of scientific problems. Expressing concepts and issues in a unique format.

Directions These directions may be used for all the poetry activities in this chapter. Students should write poems individually.

1. Discuss poetry writing. You can use Handout 7-1: Writing Poems, as a guide.
2. Introduce the specific form of poetry that you plan to discuss using Handouts 7-2 through 7-7. Each handout emphasizes a different form.
3. Write a topic on the board, or have students use topics from their daybooks.
4. Read some examples of the form of poetry you are emphasizing. You might use some you have written.
5. Lead the class in writing a class poem or verse.
6. Have students write their own poems. Remind them to use prewriting activities before writing the first draft of their poems. Brainstorming or clustering might help students generate ideas.

7. Have students write a first draft. Depending on the time allotted, this assignment can be homework.
8. Ask students to share their poems in response groups for comments.
9. Have students edit and rewrite their poems.
10. Ask volunteers to read their poems aloud to the class.
11. Collect and score poems. Save for possible publication.

Follow-Up/Assessment Poems should be scored on correct use of science concepts and adherence to form. Poems are easily shared and published. They can be entered in several different kinds of competitions. Poems are a venue for cross–grade-level sharing. Poems that express science concepts or issues are memorable. Long after your study unit is over, students will still be reading or sharing poems that they wrote.

Scoring poetry can be as simple as a check for completion, or, if you are having students write a final draft for publication, a scoring guide can be used. A language arts teacher can score the use of language component while you score the science component. A simple scoring guide might look like the one in Figure 7-1.

	Science Concepts	Use of Language
5	• demonstrates conceptual understanding • appropriately uses facts and explanations • no scientifically naïve conceptions	• focuses on concept being explained • follows poetic form in spirit if not to the letter • uses appropriate language
3	• mixes facts and explanations • shows vague conceptual understanding • no naïve conceptions	• loses focus • follows poetic form to the letter • shows skill with, but not mastery of, language use
1	• uses facts only • does not demonstrate conceptual understanding • may contain naïve conceptions	• has no focus • does not follow form • misuses language

Figure 7-1
Sample Poetry Scoring Guide

Songs

Rhythm and music have been used throughout the history of humankind to complement learning. Primitive societies recorded their customs and beliefs in songlike chants. Modern society records its customs, beliefs, history, and events in songs. Stories and traditions have been passed from generation to generation through music.

How many jingles and television advertisements are recognizable by the accompanying music? Ask your students to make a list. Can they identify the melody used in the movie, "Close Encounters of the Third Kind" and the spoof of it in a James Bond movie and in a "Lois and Clark" episode?

Quickly, write down the word encyclopedia. Did you hear a cricket singing the letters to you? Recite the alphabet. Can you do it without singing? Music is a powerful memory device that can be used for the retention of basic science concepts.

Any concept of science can be put to music. Facts, observations, and research findings can provide information to be synthesized into verses. Some teachers think that the inclusion of songs in the classroom is essential, and not only at the elementary level.

Objectives After sufficient practice, students should be able to

- write lyrics to music using science concepts, facts, explanations, and research findings
- perform songs in front of peers
- appreciate songs performed and created by their peers

Uses Studying for exams and quizzes. Summarizing. Reviewing. Memorizing.

Directions These instructions are for writing lyrics. Students may wish to choose their favorite tunes or write their own tunes. This is an individual or small group activity.

1. As a class, brainstorm a key list of topics, concepts, or other information that might be included in a song. (If you are using this activity for a test, you might like to provide your students with a "must include" list.)
2. Divide students into small groups or individuals who will be responsible for a topic or concept.

3. Students in these initial small groups can use any of the prewriting activities from Chapter 2 to get them started.

4. While students are brainstorming ideas, you might play a tape of different kinds of music—maybe show tunes, movie themes, or even rhythm only. Students might get an idea of what kind of music or rhythm to set their songs to from the music you play.

5. Plan time for groups to write and practice their songs. You can have groups provide copies of the words to their song for the rest of the class.

6. Have students perform. You may want to go outside or have the performances somewhere other than in your classroom. Invite the nearby classes. Ask students if you may record their songs for future classes.

7. Collect songs for evaluation and posterity.

Follow-Up/Assessment Songs are easy to score. A simple checklist made at the time of whole-group brainstorming could be used for scoring. A separate grade could be awarded for performance. Performances will depend on the age of your students. Songs are an acceptable mode of sharing "the facts" with others. They also provide another avenue for memorization of important information. Songs could be chronological stories of how an investigation progressed, or a summary of findings, or even a description of elements. Norman Lear wrote several "science" songs; one of his most memorable is "The Elements."

Examples Students may ask to bring instruments or audiocassette tapes to perform with. You decide what is appropriate. Instruments and tapes add to the performance, but the noise factor has to be considered. One group of senior biology students put together a band and had the whole class clapping along.

Students in a biology class were asked to write lyrics about their body systems, including the digestive, respiratory, muscular, integument, secretory, circulatory, reproductive, excretory, endocrine, and nervous systems. The beginning of one group's song is shown in Figure 7-2.

During a multigrade-level unit on the interactions between the rock and water cycles, eighth-grade students performed a "Rock-Cycle Rap" to seventh-grade students, who were studying how rocks were formed. They adapted a rock-cycle poem to their style of rap and had the seventh graders clapping along. A segment of the poem is shown in Figure 7-3.

Figure 7-2
Sample Biology
Class Song

Life. Life.

Our body is full of life.
But it takes a lot of work,
To make it through the day,
So listen to what I say,
Hey, hey, hey, HEY.
We've got ten systems
To help us through the day.

Figure 7-3
Class Rock-Cycle
Poem Adapted
to Rap

Sedimentary

What's happening above ground
where the temperature is fine
where the wind is blowing
where the water is running
where the ice is melting

Rocks got time
time to stand around
time to stand around in the weather night and day
time for the wind to blow a little away
time for water to carry a little away
time for ice to grind a little away

Rock gets smaller, moves on the wind
finds a place to settle and
finds a place to settle and sit a little while
pretty soon along comes more and starts another layer
starts another layer, whose sitting on the bottom
It gets a little heavy sitting on the bottom
lots of layers forming
forming from the sand
forming from the sediments that blow along sand

Once a lot of layers are settling down to stay
those on the bottom are scrunched along the way
the rock gets thinner more more layers sit on top.
they get cemented together and make a new
sedimentary rock

Deep in a cavern drop by drop
layers of minerals building up
building up by layers drop by drop
minerals making crystals
for us to smooth

It all takes time
for rocks to form and change
but wow what a challenge to learn all their names
find a few friends
get a bunch of rocks
compare them to ones we made
compare them to the ones we made
so quickly in our class
draw a lot of pictures with your hands.

Techniques for Primary Research

8

*S*cientists engage in primary and secondary research, and so should students, to the best of their developmental abilities. This chapter explores primary research, experimentation, and investigation. Chapter 9 looks at secondary research, which is about students researching other people's work.

Students need time to work with the tools of the scientific trade—whether it is an Erlenmeyer flask or a chromatograph, a thermometer or a computer probe, a hanging file or the Internet—so that they can use them safely and with ease. Technology is an integral part of scientific research. Students need to understand how scientific tools work. They also need to recognize the assets and limitations of these tools.

In the science classroom, the natural curiosity of students can be tapped. Imagine a classroom where small groups of students are working cooperatively or individually at different laboratory stations. They are focused, intent, and taking notes. The classroom teacher is moving from station to station, observing, commenting, and asking questions. This is a classroom where students have taken responsibility for their own learning. This is also a classroom where the teacher has built a sense of trust and respect. By the time your students have worked with the activities in this chapter, they should have the basic skills and the confidence to conduct primary research at their developmental level. The strategies in this chapter are designed to reinforce thinking processes while improving students' competence in investigative research.

Strategies in this chapter include:

- Cooperative Laboratory Reports
- Preparing Tables and Graphs
- Observations
- Forming Hypotheses
- Writing Procedures and Directions

Cooperative Laboratory Reports

By using cooperative laboratory reports, you can introduce students to a collaborative approach to laboratory work. Instead of having students write individual reports after completing a laboratory investigation, have them work in small groups to complete a group report. Students use the framework of the complete writing process—prewriting, composing, revising, and sharing/publishing—to complete their reports. Each part of the writing process is integrated with the laboratory problems students are asked to solve.

During this process, students within a group share laboratory observations and data and prepare conclusions. The report should be revised, edited, and submitted for a group score. While students probably will not do equal work, they will gain experience in collaboration on data analysis. Your job as the teacher is to act as a consultant during the sharing time.

Take a look at some of your laboratory experiments. Decide which ones would be enhanced by a collaborative discussion and conclusion. The kinds of experiments that work well for this strategy are ones where the results are always different.

Objectives After sufficient practice, students should be able to

- work in groups to write a laboratory report
- use the complete writing process as a model to produce a laboratory report

Uses Introducing laboratory safety and methods. Writing procedures. Making observations. Graphing.

Directions Students should work in pairs the first time you conduct this activity. Later, small groups of three and four students can be used. Use one of your own introductory investigations or choose one of the following questions to explore. In one problem, students explore the following question: How many drops of water can you place on the head of a penny? In the second problem, students are given a glass of water and a handful of straight pins. They are asked "How many pins can you float on the surface of the water until the water overflows?" Both of these investigations are related to surface tension. They are simple to do, do not require expensive equipment, and can provide days of practice with experimental methods.

1. Gather the materials students will need for the investigation.

2. Write the question you would like students to explore on the board. Have students copy the question into their daybooks. Allow about two minutes of quiet writing time for students to think about the question and to write down any questions or comments they have regarding the question.

3. Have students share their questions and comments with a partner. Then have pairs of students share their ideas with the class. Write the ideas on the board. Discuss.

4. Ask students what procedure they would use to answer the question. Students can work with their partners to write a simple procedure. They should also make a data table to record their findings.

5. Check students' procedures and approve those that are safe and that can be conducted in your classroom with the available materials. Have pairs of students follow their procedures and record their results. Students might ask more questions, make new observations, and record those. This cycle might take one or two days.

6. Have pairs of students write two or three general statements about what they found out by doing the investigation. These statements should be shared with the class. Discuss the findings. Then ask students to write a paragraph in their daybooks about what they know about experimentation after doing the investigation that they didn't know before doing it. Ideal responses would include: "There is more than one right answer." "Experimenting doesn't mean messing around; you have to have a plan." "Sometimes you have to do an experiment many times before you 'see' anything."

Follow-Up/Assessment Since this is a starter activity on experimental methods, have students save their work. At the end of the semester, have students compare a laboratory report with this one. These two reports would be a good addition to a portfolio.

Examples The following example tells how one teacher used these procedures in her classroom.

Cooperative Laboratory Reports

We started our unit on investigative research with the activity that asks students to find out how many drops of water you can place on the head of a penny. Students chose a laboratory partner for this activity. The materials were simple—pennies, straws or eye droppers, water, paper towels, and curiosity. Our first class discussion was a bit subdued. Students were wondering just how this activity was going to be different.

Students were given the materials and asked to investigate. They had about ten minutes before I stopped them and asked for initial findings. From these initial findings, students began to think about the variables in this very simple investigation. These variables were listed on the board. Then, as students watched each other working, they began to modify their techniques in dropping water on the penny head. Students developed a procedure and wrote it in their daybooks. They went home with questions.

The next day, students varied factors as they worked on their technique— height of dropper above the penny, speed of dropping drops, size of dropper, size of drops, purity of water, condition of pennies. Each pair of students wrote a report on their findings. On the third day of this investigation, students shared their findings with the rest of the class. The sharing was lively. Gone were the bored faces from the first day. Their conclusions—well, you'll have to conduct this with your own students. My goal was to get students thinking about every part of an investigation as a place where possibilities happen.

Preparing Tables and Graphs

Graphs and tables represent ways of showing relationships, comparisons, data collection, fluctuation of variables, changes with time, and so on. Graphs and tables are important tools used by scientists to present information precisely and efficiently in an organized manner. *Tables* are used to arrange data in columns and rows. *Graphs* often take the form of circles, bars, or lines. With all of the computer graphing programs available, it should be easy for students to create readable graphs and tables. They need to know how data is transformed into pictures, and they need practice making graphs from their own data. Students can usually interpret graphs, but they often have no idea where the information in the graph came from.

Objectives After sufficient practice, students should be able to

- read and interpret simple tables and graphs
- add information to existing tables and graphs
- identify methods of graphing that are suitable for presentation of different kinds of data and different relationships
- make graphs and tables from existing data.

Uses Organizing ideas into categories. Investigating different formats of data organization. Recording and interpreting data. Improving the ability to analyze graphs and tables.

Directions This is a large group activity. This introductory activity can be repeated as often as new graphs and tables are used in your class.

1. Place several different graphs and tables around the room.
2. Write the following question on the board: Where do graphs and tables come from? Have student free-write on this question for two to four minutes.
3. Discuss and record students' responses on the board. This might be their first discussion of the origination and development of a graph or table. Take the time to check for understanding. Have students copy the responses into their daybooks.
4. Reproduce several representative graphs and tables. If possible, use one set of data, and make several different kinds of graphs illustrating the data. (EXCEL®, Microsoft Word®, and Cricket Graph® are three computer programs that make graphs.)
5. Distribute the visuals to the class. Each student should receive only one. Some students will have the same visual, but students should not be informed of this.
6. For homework, have students write down as much as they can about the structure of their visual. Then they should study the data that is represented, and write a paragraph describing the information the visual provides.
7. In class the next day, group students who have the same visual. Have the groups share their information and write a review of the visual. The review should include positive and negative comments about the structure of the visual and about the data. The purpose is to get students to realize that data can be represented in many forms, some of which are easier to interpret than others.

8. Prepare a bulletin board with the groups' interpretations of their visuals. Add general statements about graphs and tables made by your students; these will help students internalize this important information.

Follow-Up/Assessment As soon as possible in the school year have your students make graphs from data they have collected. They should be analyzing their own graphs and the graphs of others. After completing this activity once or twice, you can assign graph and table interpretation for homework assignments, or have short quizzes with new graphs. Challenge students to see how many different kinds of graphs they can find and analyze. Display these on a bulletin board.

Examples Different tables might include information about sequencing, comparisons, change in variables, change over time, frequency distribution, ranges, and variables. Different graphs might take the form of bar, line, dot-to-dot, and circle. Have students use the examples of different kinds of graphs they have collected to present the same data in a different format.

Once students are comfortable with the analysis of simple graphs and tables, give them experience in adding information to existing graphs and tables. If students have conducted an experiment in which they made a graph of data, they can do the experiment again and add the new data to the old graph. Perhaps you have an experiment that students do every year. Save the data tables each year, and have students add their new data, making a new graph that shows both sets of data. The class can also collect data, make individual graphs, and then create a class graph.

Observations

Predicting is an important science process skill. One way that scientists predict the outcome of future experiments is by identifying repeated patterns in data and experimental results. When experimental results do not fit into an expected pattern, scientists look for experimental or instrumental errors, and/or some other anomaly to explain the unexpected results. Sometimes the unexpected leads to new discoveries.

Becoming a keen observer is important in developing the ability to recognize scientific patterns. Once students become familiar with the way scientists observe, detecting patterns becomes easier. Teachers can enhance students' ability to become better observers and better at detecting patterns. The brain is a great pattern identifier; this should be used to students' advantage in the study of science.

By gradually increasing your students' level of observation, you can help them increase their observational skills. At the first level, students use their five senses to directly observe. Most people remember their elementary experiences with their five senses. Tap into this experience with your students, using a science focus. Focused free-writing activities from Chapter 2 can be used for this purpose. Choose one sense and one place or object for the focus.

At the second level, students use tools and instruments to make observations. While younger students can use the new technologies with ease, they often do not make the transfer from the original object being observed to a video or enhanced version that is "made" by a tool. Also, tools should be investigated for their limitations and for ways in which they may distort natural observations. Most people trust technology to be accurate. This is an opportunity to question and explore the limitations of the tools that people invent and use. Help students explore science tools that help measure, record data, and probe areas of the planet and solar system where they can't physically go.

The third level of observation comes from looking at an abstract representation of data in different graphic forms. Tables and graphs may reveal patterns that other forms of recording data do not. Researchers rearrange data in the hopes of discovering new or different patterns. With practice, students can do the same thing.

Objectives After sufficient practice, students should be able to

- record observations using their senses, instruments, and scientific tools
- identify major patterns in collected data
- question data that does not fit into a recognized pattern

Uses Making predictions. Organizing material. Classifying. Analyzing data from investigations.

Directions This is a large group activity. These are general directions for introducing pattern identification. The activity can be used many times. If you are not comfortable with creating a motion pattern, use a short video clip or laser disk and set it on the repeat cycle. You will need to collect pictures showing patterns. Wall, ceiling, and floor coverings make interesting patterns. This activity lends itself to a bulletin board display. Pictures, drawings, graphs, and tables showing patterns would make an interesting and useful display.

1. Gather several different pictures that show some sort of repeating pattern, or a single pattern.
2. As students enter the classroom, start repeating directions or repeat a physical motion until all students are seated and ready to begin. Ask them to describe what you are doing. Stop your repeating activity.
3. Discuss what students observed. In this discussion, lead students to the words pattern and repeating. Work with students to develop working definitions for these words. Have students write these definitions in their daybooks.
4. Distribute the pictures you collected to groups of students.
5. Have students observe the pictures. Then have them write down their observations, identifying any patterns they observe.
6. Ask groups to share their pictures and patterns with the rest of the class. Discuss the patterns students have identified.

Follow-Up/Assessment Younger students can find and copy patterns for the class. Older students can find patterns in a habitat. You can score students' changing abilities to be careful observers and to discover by collecting several instances where they make and interpret observations. The change in their abilities can be compared during self-reflection or in a student-teacher conference.

Examples Graphs, tables, and charts are an endless source of patterns. They should be used often so that students get used to really looking at them. In other activities, students first use their own senses for making observations. Then they use a scientific instrument to observe the same objects. Students could do projects comparing the efficiency of scientific tools to enhance their own powers of observation. As you prepare activities to help students, remember that patterns are everywhere. They permeate all of the scientific disciplines. Patterns can be found in time, space, and matter. And don't overlook behavior—patterns can be found there, too.

Forming Hypotheses

Students come to your science class full of questions about the natural and physical world. They have hunches, wild guesses, and their own naïve ideas of how things work. Students might assume that these guesses are hypotheses. But are they? Suppose someone said, "It will rain at 5:00 p.m. one week from now." Would this be a hypothesis? It could be considered a hypothesis if someone tested it by waiting until that time and date to see if it rained. If it did, the hypothesis could be considered verified. But such a statement is more a guess than a hypothesis and would be a poor example of a hypothesis. The key word is *tested*. To be useful, a hypothesis, like a theory, must be capable of being tested to see whether or not evidence supports it.

The word *hypothesis* has a special meaning for scientists. A hypothesis is a statement about an observation, event, or experimental result. It is tested by experimental and investigative methods that produce evidence that either supports or refutes the hypothesis. It is from hundreds of tested and verified or refuted hypotheses that laws and theories are developed. A hypothesis need not be tested exclusively in a laboratory. Field observations, excavations, and sampling techniques are other methods used to test hypotheses.

Scientific research usually starts with a question. From the question, scientists state hypotheses about what they think the answer might be. For example, suppose a scientist asked the following question: Why doesn't this lake have any living organisms in it? The scientist developed this question after making some observations about the lake. Testable hypotheses that the scientist might develop might include the following: The lake has no food for organisms. The lake has high acidity. The lake is polluted beyond the capacity of living organisms' tolerance ranges. These statements—hypotheses—are possible answers to the original question. They also focus investigations. Observations and data collection will allow the scientist to draw conclusions about the lake. These conclusions may lead to new questions and more hypotheses. Finally, after several rounds of observation/questioning/ hypothesizing/investigation, the scientist will be able to make a final statement about the lake. Will the original question have an answer? Maybe, maybe not. But the scientist will have gathered much information about the lake that might lead to other investigations.

Forming hypotheses is basic to science. This model of seeking understanding makes up the heart of scientific inquiry.

Objectives After sufficient practice, students should be able to

- explain the meaning of *hypothesis*
- distinguish between a wild guess, a hunch, and a hypothesis
- write a hypothesis and suggest ways to test it

Uses Developing questions for investigations. Focus experimentation. Focus for field work.

Directions This activity has multigrouping components. Once you have done the introductory section, hypothesis writing can happen anytime, anywhere. These directions are for those who can find a place outdoors, but the directions are easily adapted to an indoor setting. Using a video or museum exhibit are two ways to adapt the activity for use indoors.

1. For a pretest, have students respond in their daybooks to the following question: What are scientific hypotheses and how are they different from hunches and guesses?

2. Ask students to write a paragraph or two explaining how they think scientists do their work. (This can be a homework assignment, so that students are ready to discuss their ideas when they come to class.) Discuss. From the discussion, students should identify a process of forming and testing hypotheses. The steps are (1) making observations, (2) proposing questions based on observations, (3) proposing testable explanations, (4) gathering evidence, (5) interpreting evidence, and (6) asking more questions. This would be a good time to have some scientists visit your class. Students could ask them what they investigate and how they conduct their research.

3. Have students bring writing materials to an outdoor area. Have students sit silently for several minutes and observe the environment around them. When students' speech centers are quiet, their observational senses can work more efficiently.) After students have quietly observed for as long as they can, have them record their observations.

4. Have students review their observations and write questions about them. Students may use the 6WH techniques for writing questions (Chapter 2) if they need help. Questions could also start with "I wonder . . . ."

5. In small groups, have students share their questions. One person in each group should act as recorder to write the questions on a large sheet of paper. Similar questions should be consolidated.

6. Have students make quiet observations for an additional two to four minutes, keeping their new questions in mind. They should record these additional observations.

7. For homework, students can focus on one question from the field work and write several hypotheses for the question.

8. In class, have students with similar questions share their hypotheses. Direct them to revise their hypotheses for clarity and to suggest ways they might be tested either in the field or in the classroom.

Follow-Up/Assessment This activity offers a good opportunity to distinguish between nonscientific and scientific predictions. Various pseudosciences confuse and use the ignorance of people for their own benefit. When students can "be scientific" and understand the difference, they are more likely to make better decisions on scientific issues. Later in the semester, you can ask the opening question from these directions again, then have students compare their current responses to the initial ones. Students can write a statement about their new understanding of hypotheses.

Examples Natural habitats are places where students' curiosity can transform them from passive to active learners. When students open their senses (Chapter 2) to the immediate area around them and ignore other classmates, they begin to sense the wonder of the natural world. When students observe, the keys are silence and space. Any large natural habitat will work. If you are city-bound, many of the National Geographic nature specials or other PBS programs might work. The trick is to show the program with the sound off at first. The narration will distract students from the visual. Students' responses will not be as rich as when they can use all of their senses, but it is a workable approach. Don't discount your local city environment as a place to investigate. There are many microhabitats that can be worth your students' time and energy for developing hypotheses.

Writing Procedures and Directions

Following directions and writing step-by-step procedures are skills necessary for conducting scientific inquiry. Students use many processes in writing directions, including relating, describing, classifying, synthesizing, inferring, and analyzing.

Students follow directions and procedures everyday, but many find it difficult to clearly explain to others how to complete a task. Real understanding of procedures can be shown when students are able to write usable directions for someone else.

Objectives After sufficient practice, students should be able to

- follow procedures written by others
- write step-by-step directions that others can follow

Uses Writing and using experiment directions. Conducting field work. Doing special projects. Playing games.

Directions This is a group activity. Groups of three work well.

> The Scientific Method is neither tidy nor error free. Scientific research is beset with stops and starts, unproductive ventures and some apparently aimless conjecture.
>
> *Alice F. Randall*

1. Before class begins, choose a favorite recipe and duplicate one copy for each group. Cut each recipe, or procedure, into individual steps and mix them up. Remove one step from each group's procedure. You can remove the same step from each procedure or you can vary which step you remove. Place each group's procedure in an envelope.
2. Provide each group with an envelope, a sheet of paper, and a glue stick.
3. Ask each group to order the directions into a logical procedure that would produce the proper food item.
4. After students have ordered their steps, have them glue the steps on the paper in order.
5. Post each group's procedure for the class to see. Have students compare the procedures and answer the following questions: What is the same in all procedures? What is different? Would following these directions produce _____ ? Is there anything missing in this set of directions? Hopefully students will notice the missing parts when they compare procedures. Discuss what the consequences might be if they leave out important steps in procedures.
6. Give groups additional time to rearrange their directions. Post final directions for all to see.
7. Discuss student changes. How did the groups compensate for the missing direction?
8. From the discussion, decide on critical elements of "good" directions. Have students write these elements in their daybook. You can make a large poster listing these items for reference.
9. Give students the next day's laboratory procedures cut up in the same manner as the recipe. Have them reorder it for homework. Check the procedures the next day before students begin working. Ask students to predict what might happen if any of the steps were left out.

Follow-Up/Assessment One way to assess students' ability to write and follow procedures is to have groups write and exchange experimental procedures. If each group can follow the directions of others and complete an experiment, the procedures were acceptable.

Testing recipes and eating the results is another way of showing how important precise directions are. A tablespoon instead of a teaspoon of salt or baking soda can dramatically alter the taste of a recipe.

Examples To gain practice in writing procedures or directions, you can use any of several activities. In one activity, students start by writing a goal, then write several objectives to reach that goal. Next, they outline what they will do to reach each objective.

Another approach might be to have students write procedures for a laboratory experiment. Students may be given a jumbled set of directions and asked to arrange these in a logical order. Or they may be given a map and a set of directions and asked to find a place or item in two- or three-dimensional space.

Often, in middle schools, one problem is getting students to read directions, not their ability to follow them. Teachers are used to giving directions orally and demonstrating what students will do. As a result, students are not as careful as they should be when they read directions.

Have students listen to oral directions and see if they can follow them without making mistakes. Origami lends itself to this technique, as does constructing paper airplanes.

A Long-Term Research Project Using Secondary Sources

9

*S*cientific researchers do not always conduct experiments and investigations. Often, they begin testing their hypotheses by trying to find out what others have done. They ask the same questions, write the same hypotheses, but instead of heading for the field or a laboratory, they head for a library, media center, or the Internet. Scientists engage in both primary and secondary research, and so should students, to the best of their developmental abilities.

Students need time to search out what others have done so that they can gain a perspective of the richness of the history of inquiry. They also need to discover that often scientists have several points of view on a topic, issue, or concept. When scientists gather at conferences and symposiums, they share and argue their personal points of view. Students need time to work with the research tools of the scientific trade, from hanging files, periodicals, and popular digests to the Internet. This type of project uses and develops students' abilities to synthesize information. Students have to think about the information they find, weed out the irrelevant and weave in the relevant into a series of short essays that make up each chapter of their project.

This chapter explores secondary research sources—those other than the investigator's direct observations—as a source of information. By the time your students have worked with the activities in this chapter, they should have the necessary skills and the confidence to conduct secondary research at their developmental level. The strategies in this chapter are designed to reinforce thinking processes while building students' competence in investigative research. As a product of this chapter, students prepare an extensive report. The activity example used in this chapter starts at the beginning of the school year and continues until March.

Strategies in this chapter include:

- Choosing a Topic
- Finding Information
- Using the Internet
- Organizing Information
- Finishing Student Projects
  Scoring the Final Project
  Oral Presentations
  Writing an Abstract

Choosing a Topic

One of the hardest tasks you can ask of your students is to choose one topic, issue, concept, or idea to study for a long period of time. Students can gain so many educational benefits that the effort it takes to make this type of project come alive is well worth the effort.

Objectives After sufficient practice, students should be able to

- focus on a single topic
- ask several questions that relate to a single topic
- identify several different sources of information on a single topic

Uses Starting a long-term research project. Beginning any project.

Directions You can introduce the project on the second day of school, with the intent of having students become experts on a science topic, issue, concept, or idea. In this introductory activity, students define the qualities of an "expert." Then, according to their definition, they spend over several months trying to become experts in their chosen topics. While this is an individual activity throughout, students should be encouraged to discuss their ideas with each other.

1. Duplicate and pass out Handout 9-1: Conducting Research, or a similar handout that you create, as an introductory tool. Introduce the idea of a research project where each student will be responsible for his or her own project. The project should be long-term. Decide in advance how long, and let students know the time frame on the first day of the project.
2. Ask students to fill in the first section of the handout. Using the handout, students can define what an expert is and does. These characteristics can become the basis for a scoring guide used at the end of the project.

3. After discussing these initial characteristics, ask students to think about areas in which they might like to become an expert. The list on the handout can be used to start students thinking about different fields of science. Help students identify other fields they might study. Younger students often will choose an animal or physical process; older students should be encouraged to choose more challenging topics. You decide what parameters your students can work within. If you are teaching a discipline-specific course, you may wish to have students choose a topic that parallels the course, or you may leave it wide open.

4. The filtering process can take awhile. Students need time to explore the possibilities. You can help this process along by starting the school year with a wide range of activities. You also might provide a bulletin board showing pictures of different scientists at work, plenty of global scenes, electron microscope pictures, and other images that will spark student interest.

5. Have students insert due date on the blank in the last paragraph of Handout 9-1: Conducting Research. Consider giving students a month to choose a topic. During this time, provide several classroom opportunities for students to work on their ideas.

6. For students who just can't seem to make up their minds about a specific topic, you might use the following process.

   a. Once students have chosen at least one general topic, use Handout 9-2: Brainstorming Topics, to help them narrow their ideas. Students write their possible topics on the handout and then brainstorm what they know about the topic. This worksheet could also give students the starting ideas for the chapters that will finally become their report. The focus for students' brainstorming is the list of characteristics that make an expert. You could start off students' brainstorming with a statement such as the following: To become an expert on your chosen topic, what do you have to know and be able to do?

   b. Students work in class, with partners or small groups of students who have chosen similar topics, to develop a long list of ideas and questions.

   c. Collect the completed Handout 9-2 so you can begin the record-keeping process. Set up a database to track what has been submitted and reviewed. Also, make a large master poster with students' names and topics, on which students can mark their own progress. You decide what works best for your record keeping. If students only have a few ideas on their brainstorming list, they might be encouraged to choose one of their other topics.

7. Once students have focused on a specific topic, distribute Handout 9-3: Resource Availability. Arrange a class period in your media center or library, and then allow students a week or more to search out what and where information on their topic is available. Return Handout 9-2: Brainstorming

Topics for this activity. If students are not familiar with the school's media center, enlist the media specialist to provide students with the proper instruction. This part of the project is a preliminary hunt, similar to the hypothesis-writing activity where students ask questions and then observe (Chapter 8). Instead of observing, students are seeking sources of information. If you started this activity at the beginning of the year, this part of the project should take place some time in early October. *NOTE: If students cannot find at least six different resources that have information on the topic they have chosen, they should be encouraged to go to their second or third topic choice.* You know what resources are available to your students. Set students up for success at the very beginning.

8. Collect Handout 9-3 and record progress in your chosen record-keeping system.

9. Return Handout 9-3 to students. They should keep their long-term research information in a special folder or section of their notebook. Set this up with your needs and those of your students in mind.

Follow-Up/Assessment Score Handouts 9-1, 9-2, and 9-3 for completion. From the first handout, students begin earning points for their long-term research work. If this project will last a semester or longer, students will need several visual reminders. You can dedicate one bulletin board for ongoing information about the project. Handout 9-1, where students determined the definition of an expert, can be used as a scoring checklist when students present and turn in their final projects. Check Handout 9-2 for appropriate topic choices. You might have students do a resource check on all three of their topic ideas; that way they could make a better choice about which topic to research further.

Examples Students often have a hard time choosing a single topic. To help them with this task, ask a series of questions that allow students to choose an area. The questions might lead students to make a web of their science interests. You will think of several questions of your own, but some you might consider include the following: Which current ecological issues interest you? What machines intrigue you? What new technologies are you fascinated by? Do you like to read or watch programs on space? on animals? on places? about weather? about the ocean?

Handouts 9-2 and 9-3 give students a starting place to find information. If students have at least six resources listed, they can feel confident that they will be able to find enough information to become an "expert." The "things an expert would know" that your students listed should guide their search for resources.

My students decided an "expert" would have three main characteristics: know lots of specific information, be able to answer questions, and know where and how to find out more about their topic.

Finding Information

In this part of the project, students' knowledge of finding information will be expanded and put to the test. Students will need occasional class time to gather information. Students should keep track of the information they find, using Handout 9-3: Resource Availability as an organizer. As they gather information together from many sources, they should begin to write up summary cards (see Organizing Information later in this chapter) and bibliographic information. The growth of multimedia information centers including Internet sources provides students with stacks of information. The sorting of this information is the focus of the next activity.

As students begin to reach out to others for information, they will need help in writing letters, finding addresses, using the Internet and e-mail systems, and many other tasks. If you haven't set up a collaborative network with other teachers and parents, this would be a good time to do so. There are many people with the talent to find information ready and able to help your students and you!

Objectives After sufficient practice, students should be able to

- find several different sources of information on their topics
- gather information on a specific topic

Uses Organizing information. Keeping track of work in progress and work completed.

Directions Give students one day or more a month to work in class on ongoing work. Students should be gathering and reading information for its value to their reports. This is part of the synthesis process. Students gather information and sort it according to specific topics. These topics become individual chapters.

Follow-Up/Assessment Set and keep deadlines for rough drafts. Have students keep an ongoing bibliography of their sources. Someone at your school probably has already decided what bibliography formats are acceptable for reports. Use their format for your students.

Examples Use Handout 9-4: Table of Contents as a cover sheet for each deadline in the writing process. Students should write chapter titles in pencil, because after more research, they often change them. Chapter titles reflect the sorting of information into similar categories. Students who choose to become experts on a specific organism might end up with similar chapter headings— General History, Habitat, Anatomy, Behavior, Ranges, Contact with Humans, and, in some cases, Behavior Studies or Simple Experiments. Students who choose diseases or other threats to humans might have chapter headings such as History, Threat to Humans, Current Research, and Important Researchers. Students who study historical or modern issues will have a different set of chapter titles. All headings are acceptable, as long as the information in the chapters refers to the heading.

Using the Internet

For those who are still "unplugged," research using local resources is quite acceptable and may provide information similar to that found on the Internet. If you are a "newbie" on the Internet, you can enroll the help of a "techno-surfer," one who knows how to use the Internet to find information, to get your students online. A techno-surfer may be one of your students, a parent, or a colleague. Just ask around. Most techno-surfers are willing to help out a newbie.

Students will use their ideas from Handout 9-2: Brainstorming Topics as starters for searching the Internet for information. Students may also compose short letters requesting information, and post those on bulletin boards. Key words brainstormed about their topic can be used with Web Crawler®, Yahoo®, or other Internet search engines. Lists of servers and bulletin boards should be scanned for interest groups and topics that may provide information for student projects. Depending on the sophistication of your system, students can download photos, videos, and other information. They should use their own floppy disks or sections on the hard drive to store their information.

Where do you find Web sites to get started? Many educational journals and magazines have an Internet column with the latest sites listed. Figure 9-1 lists several Web sites to get you started. Figure 9-2 lists sites for teacher resources.

Site	URL
Federal Web Locator	http://www.law.vill.edu/FedAgency/fedwebloc.html or http://www.law.vill.edu/fedagency/fedwebloc.html
Bill Nye—the Science Guy	http://www.seanet.com/Vendors/billnye/nyelabs.html
INC Pathfinder	http://www.pathfinder.com/@@F886rgAJf5/pathfinder/welcome.html
Environmental	http://www.envirolink.org
Digital Library for kids	http://www.npac.syr.edu
Busch Entertainment (animal research)	http://crusher.bev.net/education/SeaWorld/homepage.html
EE-Link (endangered species)	http://www.nceet.snre.umich.edu/Endangered.html
Yahoo (a place to start a search)	http://akebono.stanford.edu/yahoo
Knowledge, Integration, Environment (KIE) Internet Education Project (connects you to scientists, projects sites)	http://www.kie.berkeley.edu/KIE.html
Earth/space resources	http://www.rspac.ivv.nasa.gov
Zoos	http://www.mindspring.com/~zoonet
Math and Sci Elem.Ed links	http://aruba.ccit.arizona.edu/~dgifford/
Academy One	http://www.nptn.org/cyber.serv/AOneP/
Geoscience web servers	http://www.covis.nwu.edu/
UIUC-CoVis Geosciences Web Server	http://www.atmos.uiuc.edu/geosciences/geosciences.html

Figure 9-1
Selected Web Sites

Figure 9-2

Teacher Resources

Site/Description	URL
Teaching with technology	http://www.wam.umd.edu/~mlhall/teaching.html
Wentworth Communications (links to education sites)	http://www.wentworth.com
Search Engines (general)	http://www.search.com
Yahoo	http://www.yahoo.com [subject-heading search]
Lycos	http://lycos/cs.cmu.edu
Einet	http://www.einet.net/
Webcrawler	http://webcrawler.cs.washington.edu/cgibin/WebQuery
Infoseek	http://www2.infoseek.com
Web Central	http://www.tiac.net/users/thorgan/home.html
These are links to find info/ navigation links:	
Web66: a K–12 WWW project	http://web66.coled.umn.edu/
Surfwatch: blocks inappropriate sites; fee	http://www.surfwatch.com
Teachers helping teachers	http://north.pacificnet/~mandel/
K–12 activities and resources	http://www.cudenver.edu/~lsherry/
Teacher resources; student work	http://www.mcn.org/ed/cur/cw/cwhome.html
Lists of lists	http://catalog.com/vivian/interest-group-search.html

Organizing Information

In this part of the project, students' knowledge of organizing information will be expanded. Students will need occasional class time to organize information. Organizing and synthesizing information is hard work. It is more than cutting and pasting information from various sources.

Set guidelines and deadlines for turning in rough drafts of the various chapters that will make up a student report. Just like shorter essay writing, students will prewrite when they gather information from various sources. Then they write rough drafts of their chapters. They will edit with your help, and they will rewrite. This part of the process would benefit from a partnership with other teachers so that part of the composing and editing might be done under the tutelage of a language arts teacher. Otherwise, plan at least two days a month in class for writing and editing draft chapters.

Students should keep track of what they have written and its state of completion on an organization page similar to Handout 9-4: Table of Contents. Set frequent deadlines for submitting chapters. With a quick editing glance, you can make comments as students progress.

Quotable material should be copied exactly, using an appropriate citation format. Students should be warned about "direct pasting" of information for two main reasons. Besides being against the law, it is in bad scientific form to use someone else's idea without referencing them. Secondly, it is very boring to read a report that is made up of direct encyclopedia excerpts.

Objectives After sufficient practice, students should be able to

- read, sift, and incorporate information into a report.
- organize information on their topics into categories
- write short essays (chapters) on various aspects of their topics
- recognize plagiarism and take action against it

Uses Organizing information. Keeping track of work in progress and work completed. Synthesizing information into coherent chapters.

Directions Give students one day or more a month to work in class on ongoing work. Students should be gathering and reading information, writing, editing, and rewriting until a chapter is complete. Chapter titles will filter directly from students' idea sheets, depending on their individual topics. Handout 9-4 can be used as an organizer and checklist for grading. After students sort information into similar categories, they should write a summary card for each separate idea. Writing from summary cards, which may be 3 x 5 index cards or pages in a computer, changes a student who uses excerpts from resources to one who writes "expertly" about a subject. If students think of each chapter of their report as a separate essay, the whole report should be easier for them to complete.

Follow-Up/Assessment Set and keep deadlines for rough drafts. Have students keep an ongoing bibliography of their sources. Someone at your school probably has already decided what bibliography formats are acceptable for reports. Use their format for your students. Handout 9-4 gives students an ongoing organization plan for their report. Points can be given for each rough draft and final draft. A scoring guide based on students' criteria for what makes an expert can be used to score the contents of each chapter. Student peer reviews of new chapter work can ease your reading time, as well. (See Response Groups in Chapter 1.) Check rough drafts for plagiarism and "pasting" and return any samples of these to students as unacceptable.

Examples Use the table of contents page as a cover sheet for each step in the process. Students can write chapter titles in pencil because after more research, they often change chapter titles. Each chapter in this report might be written from a different perspective, as represented in the Cubing activity (Chapter 5). Students could start with a paragraph on each, then expand these paragraphs into chapter-length essays.

The pursuit of becoming an "expert," according to your students' definition, should be challenging. An expert is not one who simply excerpts the writing of others, but one who synthesizes information into a new format. Students should be able to organize the information they find into a new and unique format. An expert is one who uses the work of others to support her point of view. Students who pick issues or problems should choose a personal point of view and use the information to support their own thinking. An expert is one who makes a new claim of how something works and uses the works of others to support their ideas. Students should be able to support their points of view using evidence and facts.

Finishing Student Projects

Around February or March, students should have received letters back from their queries, spent many hours in the media center, interviewed local experts, and completed most of their written work. To help students prepare for the final written product, distribute Handout 9-5: Research Report Update, which can be used as an organization scheme for the complete report. Alter it to suit your particular needs. As a reminder, have your students write in the due date.

As part of the completion of this project, students will present an oral summation of their findings, write an abstract, and turn in their final product.

Objectives After sufficient practice, all students should be able to

- present their information orally
- summarize their findings in an abstract
- write a report using primary and secondary sources

Uses Practicing presentation methods. Celebrating the completion of a long-term project.

Directions Students will present their information as individuals, but they should be allowed to work in small groups as they prepare.

Scoring the Final Project

1. Distribute Handout 9-5: Research Report Update. It should reflect the items you think are important for the final written report. Discuss the handout with students. Use class time for students to do the final organization of their reports.
2. On the day reports are due, plan an appropriate researchers' celebration. You might have several guest speakers to congratulate your students for sticking to their individual projects. Your local paper might be interested in taking a picture of your students with their finished reports. You might also plan the celebration day as a science conference. Students would sign up for similar "sessions" (small group discussions). They would discuss their findings with each other.

Oral Presentations

1. The week before the oral presentations, conduct a class discussion in which you revisit the initial list of statements students generated about what makes an expert. Decide what should be included in the oral presentations. Set

this up as a checklist, and duplicate it as a guideline. Provide the guidelines to students ahead of time so that they know what is expected. At this time, you can prepare the presentation schedule by having students sign up for predetermined time slots. Figure 9-3: Research Scoring Guide (located at the end of this chapter) was devised by middle school students to score the projects and the oral presentations.

2. On the day of the oral presentations, duplicate a simple response form for the "audience" to fill out for each speaker. A response form might be as simple as the one in Handout 9-6: Student Response Form. After each presentation, allow a few minutes for students to reflect on what they heard. This will also give the next speaker time to set up.

3. Collect response forms and award points for completion. Cut up the response forms and sort by speaker. Give them a quick read, removing any that are not appropriate. Staple the forms for each student together and distribute them to students as feedback.

Writing an Abstract

1. During a short class discussion, have students list the characteristics of an abstract. This can be a short list on the board or a note-taking session led by you. Distribute Handout 9-7: Abstract Focus Questions. As homework, have students write rough drafts of their abstracts, which they will bring to class for peer editing and rewriting.

2. Have students work in small groups to read, review, and rewrite their abstracts.

3. Collect and score.

4. Bind abstracts together to make a class book on student projects.

Follow-Up/Assessment Figure 9-3 is an comprehensive scoring guide that could be used to score the compete project. It can be posted early in the semester as a guide for achievement. On the other hand, for the written report, a simple point checklist can be used for awarding points for completion. Handout 9-8: Research Report Scoring Sheets can be duplicated as needed. Change the division of points to suit your needs.

You might consider putting all of the class abstracts together in a book. Invite a reporter from the local newspaper to visit on presentation day to take pictures for the newspaper. Students can give the reporter information about their projects and the process they went through to produce them.

Examples One teacher put all of the abstracts in a class book and included pictures of each researcher. In fact, students spent a day as book binders. Every student report was spiral bound. Each student spent extra time making a cover and writing an About the Author page. These reports had a long life. They were presented in class orally and in a written format. Some students used their papers as an entry in the local science fair that accepted secondary research as a separate entry. Some students were asked to include their reports as part of a display at the County Office of Education. Some reports became a part of the classroom library and were used as models for the next year's class.

Figure 9-3 Research Scoring Guide

Level of Achievement	Content	Process	Presentation
Exemplary	• shows expertise for subject area according to classroom notes • written in own writing, not copied • interesting, engaging writing • shows reliability	• report is more than ten pages • ink or typed • title page present • bibliography present and in proper format • author page present • pictures present • no grammar, style, or spelling errors	• presentation is exceptional • visual aids excellent • engages audience in subject with some sort of activity
Standard	• shows expertise for subject area according to classroom notes • written in own writing, not copied • interesting, engaging writing	• report is ten or more pages long • ink or typed • title page present • bibliography present and in proper format • author page present • pictures present • few grammar, style, and spelling errors	• student shows competence with subject • answers questions • uses visual aids—laser disk, pictures, or other
Apprentice	• written in own writing, not copied • shows competence in subject area according to classroom notes • interesting writing	• report is ten pages long • in ink • title page present • bibliography present • author page present • pictures present • many grammar, style, and spelling errors	• student shows competence with subject • answers questions • doesn't use laser disk
Novice	• written in own writing, not copied • shows some competence in subject area according to classroom notes	• report is five to ten pages long • title page present • bibliography present, but may have errors • author page present • pictures present, but may be in pencil or sloppy • many grammar, style, and spelling errors	• presentation okay • no questions answered • no visual aids
No achievement	• does not turn in report	• no report	• no oral presentation

Handouts

Handout 1-1

Showcase Portfolio Directions

It's time to put your showcase portfolio together for the end of this term. It's time to start writing. Sift through the work you have done in class, and choose work that represents the following:

1. A written statement that is a general introduction in which you write about what you have learned and how you have grown (or what has prevented you from growing) in your understanding of science.

2. Your best laboratory design, with a cover letter explaining why you think it is your best. Also include in your cover letter a brief description of the kinds of experiments you would like to do next term.

3. Your best effort at research, with a cover letter describing how you did your research. In your cover letter, write about an additional place you would have liked to look for information.

4. A description, picture, or disk of a nonwritten assignment and a cover letter telling how you enjoyed this assignment.

5. One assignment that shows your ability to work in a group, with a cover letter telling how you contributed to the group and what you might do to improve your collaboration abilities.

6. Your goal statements.

7. Up to three other examples of work that show your competence and confidence in science. Remember that each additional example you add must have a cover letter explaining why you thought it was important to include in your portfolio.

- When you turn in your portfolio, remember that you need to make a current table of contents.

- Place your cover letter in front of your work.

- Place your written statement first.

Handout 1-2

Initial Goal-Setting Directions

Answer in paragraph form the following question about your abilities as a young scientist:

How am I doing in science?

Sit in a quiet place and think about yourself as a young scientist. Use the following questions to help you with your paragraph.

1. In comparison to our science classroom standards, where do I fit in?

2. What work in this science class am I proud of?

3. Where do I need to improve? (behavior, class work, homework, and so on)

4. What goals will I set for myself to accomplish between now and the next progress report period?

Examples:

Where will I improve?

How will I be the best that I can be?

How will I help others be their best?

Handout 1-3

Subsequent Goal-Setting Directions

It's time to revisit your first "How am I doing in science?" statement (Handout 1-2). Follow the directions below.

1. Reread your initial paragraph about your abilities as a scientist.

2. Review the goals you set. Did you reach those goals? If so, make new ones. If not, restate how you will achieve your current goals.

3. Write a new paragraph. Be sure you answer the questions below.

Use the following questions to help you with your new paragraph.

1. In comparison to our science classroom standards, where do I fit in?

2. What work in this science class am I proud of?

3. Where do I need to improve? (behavior, class work, homework, and so on)

4. What goals will I set for myself to accomplish between now and the next progress report period?

Examples:

Where will I improve?

How will I be the best that I can be?

How will I help others be their best?

After writing your new paragraph, attach it on top of your first paragraph and put it into your portfolio.

©Addison-Wesley Publishing Company, Inc./Published by Dale Seymour Publications®

Handout 1-4

Evaluation and Showcase Portfolio Directions

For this last marking period, you will be updating and making a Showcase Portfolio to take with you to your next year's science teacher. You can use work that is already in your portfolio, or you can add other classroom work as well. Remember, the purpose of the portfolio is to show off your work.

What's in it?

1. Your student statements. You should have six stapled together. This last one should be in your best handwriting or typed. Be sure to cover these questions in your statement.

   - What are you most proud of this year in science?
   - In what area of science do you think you have improved the most?
   - As a "student scientist," what goal(s) will you set for next year and beyond?

2. Your completed science process paper.

3. Your best lab report with a cover letter explaining why you think it is your best and how it shows your capacity to design and carry out inquiries.

4. Your favorite work—or an outline of what you did if it was a project that demonstrates your ability as a scientist. Include a cover letter explaining how the work shows you are a scientist, what you can do to improve your scientific abilities, and the direction in which you want to grow in your scientific abilities.

5. Your parent/guardian response form. You must explain your work within the portfolio and have your parent or guardian write a response to your work.

6. A Table of Contents with all work in order and a stamp of approval/completion from me.

When is it due? Portfolios with parent/guardian response form are due no later than _____ .

Please note: All cover letters and the last student statement must be typed or written in blue or black ink. All other work stored in your portfolio for this year should be removed when you take it home for your parent's response. In this portfolio is a selection of your child's work in science for this year.

©Addison-Wesley Publishing Company, Inc./Published by Dale Seymour Publications®

Handout 1-5

Parent Response Form/ Science Portfolio

Please take a few moments to look through this work with your child.

As a teacher, I am very proud of each and every student's progress in their abilities as a scientific researcher and experimenter. I think that the portfolio and daybook are fine examples of what your child has accomplished this year.

Please take a moment and write a response to your child's work. The following statements might help you focus your response.

- What I like about my child's work is . . . .
- I am amazed that my child can . . . .
- I am concerned about . . . .

_____

_____

_____

_____

_____

_____

_____

_____

_____

_____

_____

_____

_____ _____

Signed Parent/Guardian   Date

Handout 4-1

Classification of Animals

Create a classification system for the organisms shown on this sheet. Give reasons for your method of classification. Explain how your classification system could be useful to others who study these organisms.

Date _____

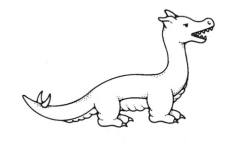

Fact or Opinion?

In the space provided for each statement, write an *F* if the statement is a fact or *O* if the statement is an opinion.

1. ____ Microscopes should be carried in an upright position.

2. ____ A binocular stereoscopic microscope has two oculars.

3. ____ ATP stands for adenosine triphosphate.

4. ____ Plants are good to eat.

5. ____ Paramecium is a single-celled organism.

6. ____ No two objects can occupy the same place at the same time.

7. ____ There are nine planets in our solar system.

8. ____ Thermometers measure the change in the temperature.

9. ____ It is hot today.

10. ____ Eagles flying 75 miles per hour have been clocked with radar guns.

11. ____ Basketball is cool.

12. ____ DNA replication occurs in the cell nucleus.

13. ____ I like cats better than dogs.

14. ____ The circumference of a circle is equal to πr^2.

15. ____ Sound waves travel slower than light waves.

16. ____ Birds lay eggs.

17. ____ Snakes have no external appendages.

18. ____ Lobster is good to eat.

Handout 4-3

Cause and Effect

Part 1 Read the following cause-and-effect relationships.

Cause	Effect
The fire alarm sounded.	Everyone left the room.
The 12:30 bell rang.	Everyone left for lunch.
The cafeteria is out of pizza.	Students will have to choose something different.
You missed the bus.	You will have to walk to school or get a ride with a friend.

Often, several causes produce a single effect, and several effects may result from a single cause. Knowing this, go back and suggest reasons for why each cause happened. Next, suggest other possible effects.

Part 2 In this section, the causes and effects are mixed up. Draw a line to match pairs of statements that fit together to illustrate a cause-and-effect relationship. Write the proper letter in front of each statement to identify which is the cause (C) and which is the effect (E).

	Thunder was heard.		There was a wind storm.
	Trees blew down.		Plant's leaves drooped.
	The light quit working.		The plant's soil was very dry.
	Lightning was seen.		The power went off.

Part 3 For each of the following statements, write a cause or an effect.

	Cause	Effect
The flask boiled over.		
The bell didn't ring.		
The light turned green.		
Wildlife died from eating oil.		
The crops were abundant this year.		
The road was icy.		
The modem would not connect.		

Comparisons

Part 1 In the table below, hydroelectric and natural gas are compared. On the back of this paper, make a similar table comparing two other forms of energy. List several characteristics of each form.

Similarities	Differences	
	Hydroelectric	*Natural Gas*
both can be transformed to other forms of energy	easily transformed to mechanical	easily transformed to heat
	renewable	nonrenewable
"harnessed" by technology		
	used wherever there is water	must be piped

Part 2 The table below compares a hurricane to a tornado. Add your ideas to the table. Include several characteristics for each item.

Similarities	Differences	
	Hurricane	*Tornado*
both are rotating wind		
speeds vary		
powerful	most powerful over water	over land only
destructive		

Handout 4-5

Comparison Question

Comparison Question: _____

_____

Part 1 List several similarities and differences about the two subjects in the question above.

Similarities	Differences

Part 2 Free-write your response to the question above. Continue on the back of this page.

©Addison-Wesley Publishing Company, Inc./Published by Dale Seymour Publications®

Persuasion Review

Use this page to review someone else's paper. Answer the following questions after this person has read you his or her paper *twice*.

Name of person whose paper you reviewed _____

1. State the issue being discussed.

2. Were the facts clear and believable?

3. If you answered "no" to question 2, which facts need clarification?

4. If you answered "yes" to question 2, which facts were presented clearly?

5. How could the sequence of presentation of the facts be improved?

6. List any unsupported opinions that the writer included.

7. Are you convinced? Explain your answer.

Weather Words

Part 1 With others on your team, classify the words below into groups. You can use a word more than once. Give each group of words a title. The title does not have to be from the list of words. Look up the definition of any word you are unsure of. Write your word groups below.

hot, warm, polar, damp, cloudy, calm, clear, foggy, cool, stormy, mild, humid, dry, tropical, wet, moist, overcast, cold, sunny, windy

Part 2 Look at each group of words you made. Rearrange the words in each group to reflect some sort of order. Write the words in the new order below. Explain the reason you placed the words in the order you did.

Practice Essay

Scientists often use charts, graphs, and tables to accompany an essay on a science topic. In this essay you are asked to examine the graphs and then write an essay based on the information given.

The bar graph shows world population growth from 1400 to 1989. The line graph shows a prediction of what the world population will be in the year 2019 if the present rate of growth continues. In your essay, compare the two graphs and include the following: (a) Describe the world population between 1400 and 1700 and explain why the population statistics showed little change. (b) Describe the world population growth from 1800 to 1989. (c) How might you account for the change in population growth from 1800 to 1989? (d) Describe what you think could alter the projected world population growth shown in the line graph.

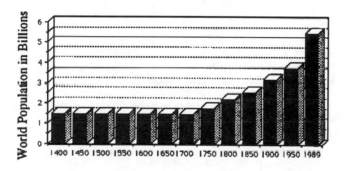

Years 1400 to 1989

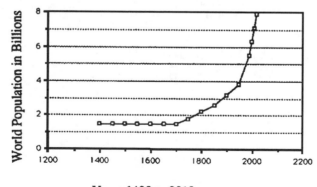

Years 1400 to 2019

Focus on Cubing

Write your topic.

Look at the illustration. Each square represents a different perspective from which you might write. Use the focusing questions for each perspective to help you write down your ideas about the topic.

Describe it.
Think of a mental picture and describe it. List dimensions, colors, name it, etc.

Associate it
What does it make you think about? Where did you first see it?

Analyze it.
What are its parts? Its uses? How is it made?

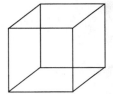

Apply it.
What do you do with it? How could it be used?

Compare it.
Write metaphors - It's like...

Argue for or against it.
Should we keep it or toss it out? Defend your position.

Write a general statement about the topic.

Handout 5-4

Cubing Response Form

Author (who reads): _____

Read through the critique questions. When your classmate reads his or her paper for the first time, listen only. During the second reading, ask the reader to clarify information for you, and make your comments in writing.

1. Write down two words that were descriptive.

2. What mental picture did the writer create for you?

3. What simile did the writer use for comparison? Can you think of another simile the writer might use?

4. With what was it associated?

5. List one way it was applied.

6. Did the writer argue FOR or AGAINST? List one supporting statement the writer used in his or her argument.

7. What do you think the main idea or general statement of this paper is?

Cubing Checklist

Provide the following information about the final-draft response sheets.

1. List the names of those who gave you a response sheet. Attach their response forms to this paper.

2. List the names of those for whom you wrote a critique.

Complete these statements:

1. I made the most improvement in (identify perspective):

2. I was the best in my critique for (fill in the person's name and the perspective):

3. A general statement/topic sentence for my essay is:

Handout 6-0

Who or What Am I? Scoring Guide

Author of Description _____

Complete the chart.

	1	2	3	4	5	6	7	8	9	10	11	12	13	14	15
Check the statements that made sense to you.															
Check the statements that you had trouble understanding.															
Check the statement number that indicates where you thought you could identify the object.															

Answer the following questions.

1. Choose one of the statements that made sense to you and explain in detail why it made sense to you.

2. Choose one of the statements you had trouble understanding and explain in detail why you had trouble.

3. What do you think is being described in this Who-or-What-Am-I description?

Writing Poems

Below are some guidelines for writing poetry. Refer to them often as you gather information for poetic writing.

- Write in the present tense.

- Observe natural objects closely. Unseen wonders may reveal themselves.

- Be patient.

- Take notes when you are in natural surroundings. Ask and answer questions about what you observe.

- Reflect on your notes in a quiet environment.

- Write about nature just as you visualize it. Choose words that suggest the season, location, or time of day (for example, *slanted sunlight; cool, salty breezes; cold northern breezes; fragrance of apple blossoms*).

- Work on each poem until it suggests exactly what you want others to sense.

Add your own guidelines here.

Handout 7-2

Bio Poem

Bio Poems describe a person, place, thing, or idea in eleven lines. The structure is set by what must be described in each line. Words in parentheses are not written, but are to be used as guides.

Format

line 1   (first name or name of term)

line 2   (four traits that describe character or term)

line 3   Relative of _____ (brother, mother, and so on)

line 4   Lover of _____ (list three things or people)

line 5   Who feels _____ (list three items)

line 6   Who needs _____ (list three items)

line 7   Who fears _____ (list three items)

line 8   Who gives _____ (list three items)

line 9   Who would like to see _____ (list three items)

line 10 Resident of _____ (town, country, imaginary place)

line 11 (last name or restatement of term)

Albert

Modest, deep-thinking, curious, compassionate.

Husband of Elsa.

Lover of Elsa, mathematics, and nature.

Who feels little for money, compassion for the underdog, and a fondness for classical music.

Who needs companionship, time to think, little praise.

Who fears Nazism, unfinished work, publicity.

Who gives support to Zionism, advice to presidents, little concern for monetary gain.

Who would like to see world government, all people free, a unified field theory.

Resident of Princeton

Einstein

Rachel

Scientist, writer, compassionate, curious.

Daughter of Maria.

Lover of the sea, nature, and purity.

Who feels passionate about protecting, exploring, and preserving the marine environment.

Who needs to write, explore, and share her scientific findings.

Who fears man's conquest of nature, pesticides, and government cover-ups.

Who gives her books to posterity, The Edge of the Sea, The Sea Around Us, and Silent Spring.

Who would like to see all creatures of the earth receive the same respect and protection.

Resident of Southport

Carson

Handout 7-3

Cinquain

A *Cinquain* (pronounced *sing-KAIN*) is a five-line poem that describes something. It usually occurs in two major formats, shown on this page. The exact formats are not stressed in the examples.

Format I

line 1 noun
line 2 two adjectives
line 3 three action verbs
line 4 four feeling verbs
line 5 a restatement of line 1

Lodestone
Strong, attraction
Grasping, holding, forceful
Opposites attract. Likes repel.
Magnet.

Format II

line 1 two-syllable subject
line 2 four syllables describing subject
line 3 six syllables showing action about subject
line 4 eight syllables showing feeling about subject
line 5 two syllables stating another word for the subject

Magma
Mellifluous
Flowing, running, active
Magma is subterranean
Hot rock.

Handout 7-4

Diamonte

A *Diamonte* is a seven-line poem that is roughly shaped like a diamond.

Format

line 1   one noun
line 2   two adjectives
line 3   three verbs ending in -ing
line 4   four nouns related to subject
line 5   three more verbs ending in -ing
line 6   two adjectives
line 7   one noun

Examples   These examples do not follow the format exactly, and you need not either. Don't feel restricted by the format.

Mammal
Upright, backbone
Flying, swimming, running
Air, land, sea, sand
Hunting, crying, dying
Tall, short
Man.

Ozone
Twin, triplet
Shielding, protecting, preserving
Responsibility, technology, duty, danger
Thinning, fading, disintegrating
French-fried
Us.

Reptile
Dry, vertebrate
Moving, crawling, swimming
Water, air, desert, earth
Walking, sliding, breathing
Cold-blooded, smooth
Snake.

Mammal
Little, strong
Kicking, punching, moving
Water, air, nutrients, blood
Breathing, living, growing
Helpless, defenseless
Baby.

Handout 7-5

Haiku

Haiku is a form of Japanese poetry. Each poem is made up of three lines and seventeen syllables. Each is meant to express a picture of nature. The poem should not rhyme, but should flow easily from the first to the third line. The correct number of syllables is desirable, but not necessary, for beginners.

Format

line 1   five syllables
line 2   seven syllables
line 3   five syllables

Examples

Windy day in May
Changing without a notice
Blowing us about.

Even my dog
hesitates at the doorway
newly fallen snow.

Summer thunderstorm
nature's way of brightening
a lonely journey.

Hanging upside down
the caterpillar spins a
sun-sparkled cocoon.

Write your haiku here.

Sonnet

Sonnets are composed of fourteen lines, with rhyming words at the end of every other line. Beginners need not follow this rule exactly. It's more important to be expressive about your subject.

Examples

Tropical rain forests will soon be gone,

and with them all the birds so bright.

Never more to hear their song

or strange animals to see through
    filtered light.

The trees are burned so cattle graze

to serve our fast-food chains

with a quick-made hamburger craze.

Some call it progress on the range,

without knowing the land won't last.

For poor soil there is the norm

so more trees must be slashed

believing more jobs will be born.

Such futility escapes are reason

there won't be jobs with all the
    trees gone.

Tropical rain forests hides answers many,

for all sorts of questions but no time
    to ask,

about cures for diseases where there
    aren't any,

known at present, leaving the task

of trying hard to save these trees and
    jungle.

There live plants and animals plenty,

as yet unknown and yet we bungle,

a chance to know some answers to

riddles of nature yet unknown.

If only there was more time to ask

who, these people are that slash and burn,

without giving science a chance to learn

the secrets hidden with time running out

to learn what rain forest life is all about.

Handout 7-7

Limerick

A *limerick* is a poem of five lines, usually with the rhyme scheme *aabba*. The first, second, and fifth lines have three stresses (*feet* or *meter*), and the third and fourth lines have two stresses. As with other forms of poetry, you need not follow the format exactly. The fun part about limericks is the meter. The form was popularized by Edward Lear.

Examples

There once was a botanist named Linnaeus,
Who sought to bring order to chaos.
Who said to claim it,
You have to name it.
Hoorah for the good Swede, Linnaeus.

Charles Darwin was involved in detection,
Of every botanical and zoological section.
He studied finches and trees,
Mammals and bees,
And theorized about natural selection.

Madame Curie studied physics most furious,
Glowing pitchblende made her so curious.
Radium she extracted,
Radiation poisoning she contracted,
Made twice Nobel Prize-winning glorious.

Conducting Research

Your challenge is to become an expert on a particular science topic. What is an expert? What are some characteristics of an expert? Write your ideas below.

You may choose from several different areas of science. Some areas that you might investigate include

- living organisms
- nonliving things
- science concepts
- space exploration
- heavenly bodies
- science phenomenon
- science personalities
- history of a law or theory

List below any additional areas of science that you might like to investigate.

We will investigate the library as a source of information about science topics. You should look around the library and begin to single out your research topic. By _____ , you must have chosen your topic. Earlier submissions are permitted. Choose wisely.

Brainstorming Topics

Choose three general topics that you would like to research.

Topic Idea 1 _____

Topic Idea 2 _____

Topic Idea 3 _____

After you have chosen your topic ideas, brainstorm a list of ideas and words that relate to each of your topic ideas. Write your ideas below.

Resource Availability

Write your topic idea: _____

Indicate those places where you found information that might be useful when you start your research. List specific information about the source in the description column.

Check (✓)	Source	Description
	Books	
	Businesses	
	Card catalog	
	Encyclopedias	
	Experts	
	Film strips	
	Government agencies	
	Internet sites	
	Journals	
	Films	
	Newspaper articles	
	Organizations	
	Posters	
	Video disks	
	Videos	

Table of Contents

Topic: _____

Chapter	Chapter Title	Date rough draft checked	Date update draft checked	Date completed
1				
2				
3				
4				
5				
6				

Notes:

Handout 9-5

Research Report Update

Deadlines are coming up soon for your report!

Your next update, which should be your nearly completed report, is due _____

The last day to turn in your complete report is _____

Use these guidelines when organizing your report:

- Include a cover page and a table of contents.

- Each chapter should have a chapter introduction page.

- Each photo, picture, or map should be on its own page and labeled.

- All written pages must be in ink, typed, or word processed. If the paper is typed, double or triple space it. Allow at least 1-inch margins. Use at least a 12-point font.

- Include a page about the author—that's you.

- Include any materials you collected for your report as appendixes—copied pages, letter responses, newspaper articles, and so on.

- Include a reference page telling where you found your information. This information should be in alphabetical order. Use the formats below.

Format for a book Author's last name, author's first name. Title of book, publisher of book, page numbers you used, date of publication.

Format for a magazine or journal Author's last name, author's first name. Title of article, name of magazine, page numbers you used, month and year of publication.

Student Response Form

Name of presenter _____

Write down the most interesting part of the presentation.

_____

_____

_____

(cut here) -

Name of presenter _____

Write down the most interesting part of the presentation.

_____

_____

_____

(cut here) -

Name of presenter _____

Write down the most interesting part of the presentation.

_____

_____

_____

(cut here) -

Name of presenter _____

Write down the most interesting part of the presentation.

_____

_____

_____

Abstract Focus Questions

Write a short summary about your research paper.

Think about how you did your report. Then answer the following questions in a single summary paragraph.

1. What did you learn about doing research?

2. What was the hardest part of doing your report?

3. What was the easiest?

4. What was the most interesting part of doing your report?

5. If you were starting your report over, what would you do differently?

· ·

Handout 9-8

Research Report Scoring Sheet

Cover (10) _____

Comments:

Table of Contents (20) _____

Comments:

Chapters and pictures (20) _____

Comments:

Bibliography (20) _____

Comments:

Author Page (10) _____

Comments:

Final Score (out of 80) _____

Comments:

Appendixes ·······························

Time and Activity Planner

Do you find yourself in a hurry and need a short activity to complete your lesson? Are you planning for a substitute teacher? This index will help you find the fillers you need. "Beginning of the Year or Unit" and "Do Me Twice—or Maybe Thrice—in a Quarter or Term" are meant as occasional activities. "Quickies" are just that—short writing activities that may be put to several uses (for example, pre- and post-tests, follow-ups, or means of evaluating a new teaching strategy). Make sure your students are clear on the writing directions. "Fifteen to Thirty Minutes" and "About an Hour" can be used for writing during science. If your students have experience with the style of writing, these activities work especially well for substitutes. "Ongoing" and "Send It Home" refer to writing that may take several sessions to complete. "Bulletin Boards" can grow from just a title or picture, using any of the listed activities.

Reports, p. 120; Forming Hypotheses, p. 126; Writing Procedures and Directions, p. 129

Ongoing

Brainstorming, p. 26; Sensory Perceptions, examples, sense of touch, p. 37; Science Daybooks, p. 42; Note Taking, Note Making, p. 45; Reading Logs, p. 47; Approaches to Problem Solving, p. 67; Concept Mapping, p. 76; Cubing, p. 87; Word Pictures, p. 97; News Briefs, p. 98; Letters, p. 99; Posters, p. 101; Comics, p. 105; Skits, p. 106; Observations, p. 124; Finding Information, p. 137; Using the Internet, p. 138; Organizing Information, p. 141; Finishing Student Projects, p. 143

Send It Home

Questioning, p. 31; Fact and Opinion, p. 56; Cause and Effect, p. 58; Comparing Characteristics, p. 62; Gentle Persuasion, p. 63; Shades of Meaning, p. 74; Fact and Fiction, p. 79; Open-Ended Essays, p. 81; Literary Devices, p. 92; Science Acronyms, p. 94; Word Pictures, p. 97; News Briefs, p. 98; Letters, p. 99; Pre- and Post-Unit Letters (Letters, Examples), p. 100; Posters, p. 101; Who or What Am I?, p. 102; Comics, p. 105; Skits, p. 106; Stories, p. 108; Poems, p. 112; Songs, p. 114; Cooperative Laboratory Reports, p. 120; Preparing Tables and Graphs, p. 122; Forming Hypotheses, p. 126; Writing Procedures and Directions, p. 129; Finding Information, p. 137; Using the Internet, p. 138; Organizing Information, p. 141

Index 2

Grouping

Students like to work in groups, but not all the time. "On My Own" activities, which often follow group instruction, allow students to think and write quietly. Collaboration, however, can lead to interesting ideas. Small groups are best when a variety of opinions and ideas are desirable. Large groups are mainly used for discussions and instruction.

Bulletin Boards

Questioning, p. 31; Shades of Meaning, p. 74; Fact and Fiction, p. 79; Literary Devices, p. 92; Science Acronyms, p. 94; Word Pictures, p. 97; Posters, p. 101; Who or What Am I?, p. 102; Comics, p. 105; Poems, p. 112; Songs, p. 114; Preparing Tables and Graphs, p. 122; Observations, p. 124; Forming Hypotheses, p. 126; Writing Procedures and Directions, p. 129

On My Own

Clustering, p. 28; Focused Free Writing, p. 33; Exit Slip (part of focused free writing), p. 34; Sensory Perceptions, examples, sense of touch, p. 37; Science Daybooks, p. 42; Note Taking, Note Making, p. 45; Reading Logs, p. 47; Instant Versions, p. 49; Summaries, p. 50; Outlines (Summaries, examples, last paragraph), p. 51; Classification, p. 54; Fact and Opinion, p. 56; Cause and Effect, p. 58; Comparison, p. 61; Comparing Characteristics, p. 62; Gentle Persuasion, p. 63; Shades of Meaning, p. 74; Concept Mapping, p. 76; Fact and Fiction, p. 79; Open-Ended Essays, p. 81; Cubing, p. 87; Literary Devices, p. 92; Science Acronyms, p. 94; Word Pictures, p. 97; News Briefs,

*Science
Processes
and Skills*

In this index, many activities show up in multiple categories. That is because they are meant to reinforce the lower-order skills as students work on higher-order skills. Work your way through the list, emphasizing each process and skill as you use a variety of activities to reinforce learning.

Observing

Communicating

Comparing

Ordering

Categorizing

Relating

p. 74; Concept Mapping, p. 76; Fact and Fiction, p. 79; Open-Ended Essays, p. 81; Cubing, p. 87; Literary Devices, p. 92; Science Acronyms, p. 94; Word Pictures, p. 97; News Briefs, p. 98; Letters, p. 99; Pre- and Post-Unit Letters (Letters, Examples), p. 100; Posters, p. 101; Who or What Am I?, p. 102; Comics, p. 105; Skits, p. 106; Stories, p. 108; Poems, p. 112; Songs, p. 114; Cooperative Laboratory Reports, p. 120; Preparing Tables and Graphs, p. 122; Observations, p. 124; Forming Hypotheses, p. 126; Organizing Information, p. 141

Inferring

Summaries, p. 50; Cause and Effect, p. 58; Approaches to Problem Solving, p. 67; Concept Mapping, p. 76; Fact and Fiction, p. 79; Open-Ended Essays, p. 81; Cubing, p. 87; Word Pictures, p. 97; Pre- and Post-Unit Letters (Letters, Examples), p. 100; Posters, p. 101; Skits, p. 106; Stories, p. 108; Poems, p. 112; Songs, p. 114; Forming Hypotheses, p. 126

Applying

Gentle Persuasion, p. 63; Approaches to Problem Solving, p. 67; Concept Mapping, p. 76; Fact and Fiction, p. 79; Open-Ended Essays, p. 81; Cubing, p. 87; Word Pictures, p. 97; Letters, p. 99; Pre- and Post-Unit Letters (Letters, Examples), p. 100; Posters, p. 101; Comics, p. 105; Skits, p. 106; Stories, p. 108; Poems, p. 112; Songs, p. 114; Preparing Tables and Graphs, p. 122; Forming Hypotheses, p. 126; Writing Procedures and Directions, p. 129; Finishing Student Projects, p. 143

Questioning

Questioning, p. 31; Science Daybooks, p. 42; Note Taking, Note Making, p. 45; Reading Logs, p. 47; Approaches to Problem Solving, p. 67; Who or What Am I?, p. 102; Stories, p. 108; Observations, p. 124; Forming Hypotheses, p. 126; Choosing a Topic, p. 134; Finding Information, p. 137; Using the Internet, p. 138

Investigating

Sensory Perceptions, examples, sound hunt, p. 37; Sensory Perceptions, examples, sense of touch, p. 37; Reading Logs, p. 47; Gentle Persuasion, p. 63; Approaches to Problem Solving, p. 67; News Briefs, p. 98; Cooperative Laboratory Reports, p. 120; Choosing a Topic, p. 134; Finding Information, p. 137; Using the Internet, p. 138

Data Gathering

Exam Review (part of questioning), paragraph before Fig. 2-5, beginning *To prepare for a final exam,* p. 32; Sensory Perceptions, p. 36; Sensory Perceptions, examples, sense of smell, p. 37; Sensory Perceptions, examples, sound hunt, p. 37; Sensory Perceptions, examples, sense of touch, p. 37; Science Daybooks, p. 42; Note Taking, Note Making, p. 45; Approaches to Problem Solving, p. 67; Cooperative Laboratory Reports, p. 120; Choosing a Topic, p. 134; Finding Information, p. 137; Using the Internet, p. 138

Explaining

Exam Review (part of questioning), paragraph before Fig. 2-5, beginning *To prepare for a final exam,* p. 32; Science Daybooks, p. 42; Note Taking, Note Making, p. 45;

Instant Versions, p. 49; Summaries, p. 50; Outlines (Summaries, examples, last paragraph), p. 51; Cause and Effect, p. 58; Comparison, p. 61; Comparing Characteristics, p. 62; Gentle Persuasion, p. 63; Approaches to Problem Solving, p. 67; Shades of Meaning, p. 74; Concept Mapping, p. 76; Fact and Fiction, p. 79; Open-Ended Essays, p. 81; Cubing, p. 87; Literary Devices, p. 92; Science Acronyms, p. 94; Word Pictures, p. 97; News Briefs, p. 98; Letters, p. 99; Pre- and Post-Unit Letters (Letters, Examples), p. 100; Posters, p. 101; Who or What Am I?, p. 102; Comics, p. 105; Skits, p. 106; Stories, p. 108; Poems, p. 112; Songs, p. 114; Cooperative Laboratory Reports, p. 120; Preparing Tables and Graphs, p. 122; Observations, p. 124; Writing Procedures and Directions, p. 129; Organizing Information, p. 141; Finishing Student Projects, p. 143

Index 4

Writing Process

Prewriting

Brainstorming, p. 26; Clustering, p. 28; Questioning, p. 31; Exam Review (part of questioning), paragraph before Fig. 2-5, beginning *To prepare for a final exam,* p. 32; Focused Free Writing, p. 33; Sensory Perceptions, p. 36; Sensory Perceptions, examples, sense of smell, p. 37; Sensory Perceptions, examples, sound hunt, p. 37; Sensory Perceptions, examples, sense of touch, p. 37; Guided Imagery, p. 38; Science Daybooks, p. 42; Note Taking, Note Making, p. 45; Reading Logs, p. 47; Instant Versions, p. 49; Classification, p. 54; Fact and Opinion, p. 56; Cause and Effect, p. 58; Comparison, p. 61; Comparing Characteristics, p. 62; Gentle Persuasion, p. 63; Approaches to Problem Solving, p. 67; Concept Mapping, p. 76; Fact and Fiction, p. 79; Open-Ended Essays, p. 81; Cubing, p. 87; Literary Devices, p. 92; Science Acronyms, p. 94; Word Pictures, p. 97; News Briefs, p. 98; Letters, p. 99; Pre- and Post-Unit Letters (Letters, Examples), p. 100; Posters, p. 101; Who or What Am I?, p. 102; Comics, p. 105; Skits, p. 106; Stories, p. 108; Poems, p. 112; Songs, p. 114; Cooperative Laboratory Reports, p. 120; Preparing Tables and Graphs, p. 122; Observations, p. 124; Forming Hypotheses, p. 126; Writing Procedures and Directions, p. 129; Choosing a Topic, p. 134; Finding Information, p. 137; Using the Internet, p. 138; Finishing Student Projects, p. 143

Composing

Exam Review (part of questioning), paragraph before Fig. 2-5, beginning *To prepare for a final exam,* p. 32; Focused Free Writing, p. 33; Admit Slip (part of focused free writing), p. 34; Exit Slip (part of focused free writing), p. 34; Science Daybooks, p. 42; Note Taking, Note Making, p. 45; Reading Logs, p. 47; Summaries, p. 50; Outlines (Summaries, examples, last paragraph), p. 51; Cause and Effect, p. 58; Comparison, p. 61; Comparing Characteristics, p. 62; Gentle Persuasion, p. 63; Approaches to Problem Solving, p. 67; Shades of Meaning, p. 74; Concept Mapping, p. 76; Fact and Fiction, p. 79; Open-Ended Essays, p. 81; Cubing, p. 87; Literary Devices, p. 92; Science Acronyms, p. 94; Word Pictures, p. 97; News Briefs, p. 98; Letters, p. 99; Pre- and Post-Unit Letters (Letters, Examples), p. 100; Posters, p. 101; Who or What Am I?, p. 102; Comics, p. 105; Skits, p. 106; Stories, p. 108; Poems, p. 112; Songs, p. 114; Cooperative Laboratory Reports, p. 120; Preparing Tables and Graphs, p. 122;

Observations, p. 124; Forming Hypotheses, p. 126; Writing Procedures and Directions, p. 129; Organizing Information, p. 141; Finishing Student Projects, p. 143

Editing

Science Daybooks, p. 42; Gentle Persuasion, p. 63; Approaches to Problem Solving, p. 67; Fact and Fiction, p. 79; Open-Ended Essays, p. 81; Cubing, p. 87; Science Acronyms, p. 94; Word Pictures, p. 97; News Briefs, p. 98; Letters, p. 99; Pre- and Post-Unit Letters (Letters, Examples), p. 100; Posters, p. 101; Who or What Am I?, p. 102; Comics, p. 105; Skits, p. 106; Stories, p. 108; Poems, p. 112; Songs, p. 114; Cooperative Laboratory Reports, p. 120; Preparing Tables and Graphs, p. 122; Writing Procedures and Directions, p. 129; Organizing Information, p. 141; Finishing Student Projects, p. 143

Sharing/Publishing

Gentle Persuasion, p. 63; Approaches to Problem Solving, p. 67; Shades of Meaning, p. 74; Concept Mapping, p. 76; Fact and Fiction, p. 79; Open-Ended Essays, p. 81; Cubing, p. 87; Science Acronyms, p. 94; Word Pictures, p. 97; News Briefs, p. 98; Letters, p. 99; Pre- and Post-Unit Letters (Letters, Examples), p. 100; Posters, p. 101; Who or What Am I?, p. 102; Comics, p. 105; Skits, p. 106; Stories, p. 108; Poems, p. 112; Songs, p. 114; Cooperative Laboratory Reports, p. 120; Preparing Tables and Graphs, p. 122; Writing Procedures and Directions, p. 129; Finishing Student Projects, p. 143

Selected Bibliography

Applebee, A. 1981. *Writing in the Secondary School: English and the Content Areas.* Urbana, IL: National Council of Teachers of English.

Associates, P. 1980. *Search for Solutions: Patterns* [video]. New York: Playback.

Brooks, J. G., and M. G. Brooks. 1993. *In Search of Understanding: The Case for Constructivist Classrooms.* Alexandria, CA: Association for Supervision and Curriculum Development.

Caine, R. N., and G. Caine. 1991. *Making Connections: Teaching and the Human Brain.* Menlo Park, CA: Addison-Wesley Publishing Company, Inc.

Elbow, P. 1981. *Writing with Power: Techniques for Mastering the Writing Process.* New York: Oxford University Press.

Emig, J. 1971. *The Composing Process of Twelfth Graders.* Urbana, IL: National Council of Teachers of English.

Freedman, R. 1987. *Connections: Writing and Science Achievement.* Middletown, PA: Pennsylvania State University.

———— 1990. *Connections: Science by Writing.* Paradise, CA: Serin House Publishers.

———— 1993. *"Writing, Student Portfolios, and Authentic Assessment in Science."* Portfolio News, 4(2), 6,17. Portfolio Assessment Clearinghouse.

Freedman, R. L. H. 1994a. *"Elizabeth Stage."* California Classroom Science, 6(3).

———— 1994b. *Open-ended Questioning: A Handbook for Educators.* Menlo Park, CA: Addison-Wesley Publishing Company, Inc..

Gardner, H. 1982. *Frames of Mind.* New York: Basic Books.

Gere, A., and R. Abbott. 1985. *Talking About Writing: The Language of Writing Groups.* Research in the Teaching of English, 19(4), 362–381.

Graves, D. 1983. *Writing: Teachers and Children at Work.* Portsmouth, MA: H. H. Heinemann Educational Books.

Logsdon, T. 1993. *Breaking Through: Creative Problem Solving Using Six Successful Strategies.* Reading, MA: Addison-Wesley Publishing Company, Inc.

Marks-Tarlow, T. 1996. *Creativity Inside Out: Learning Through Multiple Intelligences.* Menlo Park, CA: Addison-Wesley Publishing Company,

National Research Council, ed. 1996. *National Science Education Standards.* Washington, DC: National Academy Press.

Sassaman, S. 1988. *Patterns for I.D.E.A.S.* Paper presented at the Personal Notes, Middletown, PA.

Shoemaker, E. 1986. *Fort Bragg Writing Project.* Fort Bragg, CA: Fort Bragg Unified School District.

Sanford, G. 1979. *How to Handle the Paper Load*. Urbana, IL: National Council of Teachers of English.

Tierney, R. 1987. *Science and Writing*. Paper presented at the Capitol Area Writing Project Summer Institute, Middletown, PA.

Yelverton, B. and **D. Grange. 1983.** *"Crystallizing the Cinquain."* The Science Teacher, 50(1), 39. Washington, DC: National Science Teachers Association.

Young, A. and **T. Fulwiler,** eds. **(1986).** *Writing Across the Disciplines,* Research into Practice. Upper Montclair, NJ: Boynton/Cook.